Nadia Nedjah, Luiza de Macedo Mourelle (Eds.)

Swarm Intelligent Systems

T0181802

Studies in Computational Intelligence, Volume 26

Editor-in-chief
Prof. Janusz Kacprzyk
Systems Research Institute
Polish Academy of Sciences
ul. Newelska 6
01-447 Warsaw
Poland
E-mail: kacprzyk@ibspan.waw.pl

Nadia Nedjah
Luiza de Macedo Mourelle

Swarm Intelligent Systems

With 65 Figures and 34 Tables

 Springer

Dr. Nadia Nedjah
Department of Electronics Engineering and Telecommunications - DETEL
Faculty of Engineering - FEN
State University of Rio de Janeiro - UERJ
Rua São Francisco Xavier, 524, 5o. andar
Maracanã, CEP 20559-900
Rio de Janeiro, RJ
Brazil
E-mail: nadia@eng.uerj.br

Dr. Luiza de Macedo Mourelle
Department of System Engineering and Computation - DESC
Faculty of Engineering - FEN
State University of Rio de Janeiro - UERJ
Rua São Francisco Xavier, 524, 5o. andar
Maracanã, CEP 20559-900
Rio de Janeiro, RJ
Brazil
E-mail: ldmm@eng.uerj.br

ISSN print edition: 1860-949X
ISSN electronic edition: 1860-9503
ISBN 978-3-642-07041-9   e-ISBN 978-3-540-33869-7

Springer is a part of Springer Science+Business Media
springer.com
© Springer-Verlag Berlin Heidelberg 2006
Softcover reprint of the hardcover 1st edition 2006

Preface

Swarm intelligence is an innovative computational way to solving hard problems. This discipline is inspired by the behavior of social insects such as fish schools and bird flocks and colonies of ants, termites, bees and wasps. In general, this is done by mimicking the behavior of the biological creatures within their swarms and colonies.

Particle swarm optimization, also commonly known as PSO, mimics the behavior of a swarm of insects or a school of fish. If one of the particle discovers a good path to food the rest of the swarm will be able to follow instantly even if they are far away in the swarm. Swarm behavior is modeled by particles in multidimensional space that have two characteristics: a position and a velocity. These particles wander around the hyperspace and remember the best position that they have discovered. They communicate good positions to each other and adjust their own position and velocity based on these good positions.

The ant colony optimization, commonly known as ACO, is a probabilistic technique for solving computational hard problems which can be reduced to finding optimal paths. ACO is inspired by the behavior of ants in finding short paths from the colony nest to the food place. Ants have small brains and bad vision yet they use great search strategy. Initially, real ants wander randomly to find food. They return to their colony while laying down pheromone trails. If other ants find such a path, they are likely to follow the trail with some pheromone and deposit more pheromone if they eventually find food.

Instead of designing complex and centralized systems, nowadays designers rather prefer to work with many small and autonomous agents. Each agent may prescribe to a global strategy. An agent acts on the simplest of rules. The many agents co-operating within the system can solve very complex problems with a minimal design effort. In General, multi-agent systems that use some swarm intelligence are said to be swarm intelligent systems. They are mostly used as search engines and optimization tools.

The goal of this volume has been to offer a wide spectrum of sample works developed in leading research throughout the world about innovative methodologies of swarm intelligence and foundations of engineering swarm intelligent systems as well as application and interesting experiences using the particle swarm optimisation, which is at the heart of computational intelligence. The book should be useful both for beginners and experienced researchers in the field of computational intelligence.

Part I: Methodologies Based on Particle Swarm Intelligence

In Chapter 1, which is entitled *Swarm Intelligence: Foundations, Perspectives and Applications*, the authors introduce some of the theoretical foundations of swarm intelligence. They focus on the design and implementation of the Particle Swarm Optimization (PSO) and Ant Colony Optimization (ACO) algorithms for various types of function optimization problems, real world applications and data mining.

In Chapter 2, which is entitled *Waves of Swarm Particles (WoSP)*, the author introduce an adaption of the conventional particle swarm algorithm that converts the behaviour from the conventional search and converge to an endless cycle of search, converge and then diverge to carry on searching. After introducing this new waves of swarm particles (WoSP) algorithm, The author present its behaviour on a number of problem spaces is presented. The simpler of these problem spaces have been chosen to explore the parameters of the new algorithm, but the last problem spaces have been chosen to show the remarkable performance of the algorithm on highly deceptive multi dimensional problem spaces with extreme numbers of local optima.

In Chapter 3, which is entitled *Grammatical Swarm: A variable-length Particle Swarm Algorithm*, the authors examine a variable-length Particle Swarm Algorithm for Social Programming. The Grammatical Swarm algorithm is a form of Social Programming as it uses Particle Swarm Optimisation, a social swarm algorithm, for the automatic generation of programs. The authors extend earlier work on a fixed-length incarnation of Grammatical Swarm, where each individual particle represents choices of program construction rules, where these rules are specified using a Backus-Naur Form grammar. The authors select benchmark problems from the field of Genetic Programming and compare their performance to that of fixed-length Grammatical Swarm and of Grammatical Evolution. They claim that it is possible to successfully generate programs using a variable-length Particle Swarm Algorithm, however, based on the problems analysed they recommend to exploit the simpler bounded Grammatical Swarm.

In Chapter 4, which is entitled *SWARMs of Self-Organizing Polymorphic Agents*, the authors describe a SWARM simulation of a distributed approach to fault mitigation within a large-scale data acquisition system for BTeV, a particle accelerator-based High Energy Physics experiment currently under development at Fermi National Accelerator Laboratory. Incoming data is expected to arrive at a rate of over 1 terabyte every second, distributed across 2500 digital signal processors. Through simulation results, the authors show how lightweight polymorphic agents embedded within the individual processors use game theory to adapt roles based on the changing needs of the environment. They also provide details about SWARM architecture and implementation methodologies.

Part II: Experiences Using Particle Swarm Intelligence

In Chapter 5, which is entitled *Swarm Intelligence — Searchers, Cleaners and Hunters*, the authors examine the concept of swarm intelligence through three examples of complex problems which are solved by a decentralized swarm of simple agents. The protocols employed by these agents are presented, as well as various analytic results for their performance and for the problems in general. The problems examined are the problem of finding patterns within physical graphs (e.g. *k-cliques*), the *dynamic cooperative cleaners* problem, and a problem concerning a swarm of UAVs (unmanned air vehicles), hunting an evading target (or targets).

In Chapter 6, which is entitled *Ant Colony Optimisation for Fast Modular Exponentiation using the Sliding Window Method*, the authors exploit the ant colony strategy to finding optimal addition sequences that allow one to perform the pre-computations in window-based methods with a minimal number of modular multiplications. The authors claim that this improves the efficiency of modular exponentiation. The author compare the addition sequences obtained by the ant colony optimisation to those obtained using Brun's algorithm.

In Chapter 7, which is entitled *Particle Swarm for Fuzzy Models Identification*, the authors present the use of Particle Swarm Optimization (PSO) algorithm for building optimal fuzzy models from the available data. The authors also present the results based on selection based PSO variant with lifetime parameter that has been used for identification of fuzzy models. The fuzzy model identification procedure using PSO as an optimization engine has been implemented as a Matlab toolbox and is presented in the next chapter. The simulation results presented in this chapter have been obtained through this toolbox. The toolbox has been hosted on `SourceForge.net`, which is the world's largest development and download repository of open-source code and applications.

In Chapter 8, which is entitled *A Matlab Implementation of Swarm Intelligence based Methodology for Identification of Optimized Fuzzy Models*, the authors describe the implementation of the fuzzy model identification procedure (see Chapter 7) using PSO as an optimization engine. This toolbox provides the features to generate Mamdani and Singleton fuzzy models from the available data. The authors claim that this toolbox can serve as a valuable reference to the swarm intelligence community and others and help them in designing fuzzy models for their respective applications quickly.

We are very much grateful to the authors of this volume and to the reviewers for their tremendous service by critically reviewing the chapters. The editors would also like to thank Prof. Janusz Kacprzyk, the editor-in-chief of the Studies in Computational Intelligence Book Series and Dr. Thomas Ditzinger from Springer-Verlag, Germany for their editorial assistance and excellent collaboration to produce this scientific work. We hope that the reader will share our excitement on this volume and will find it useful.

March 2006

Nadia Nedjah
Luiza M. Mourelle

State University of Rio de Janeiro
Brazil

Contents

Part II Experiences Using Particle Swarm Intelligence

5 Swarm Intelligence — Searchers, Cleaners and Hunters

Yaniv Altshuler, Vladimir Yanovsky, Israel A. Wagner, Alfred M.

8 A Matlab Implementation of Swarm Intelligence based Methodology for Identification of Optimized Fuzzy Models

List of Figures

List of Tables

Part I

Methodologies Based on Particle Swarm
Intelligence

1

Swarm Intelligence: Foundations, Perspectives and Applications

Ajith Abraham[1], He Guo[2], and Hongbo Liu[2]

[1] IITA Professorship Program, School of Computer Science and Engineering, Chung-Ang University, Seoul, 156-756, Korea. ajith.abraham@ieee.org, http://www.softcomputing.net

[2] Department of Computer Science, Dalian University of Technology, Dalian, 116023, China. {guohe,lhb}@dlut.edu.cn

This chapter introduces some of the theoretical foundations of swarm intelligence. We focus on the design and implementation of the Particle Swarm Optimization (PSO) and Ant Colony Optimization (ACO) algorithms for various types of function optimization problems, real world applications and data mining. Results are analyzed, discussed and their potentials are illustrated.

1.1 Introduction

Swarm Intelligence (SI) is an innovative distributed intelligent paradigm for solving optimization problems that originally took its inspiration from the biological examples by swarming, flocking and herding phenomena in vertebrates.

Particle Swarm Optimization (PSO) incorporates swarming behaviors observed in flocks of birds, schools of fish, or swarms of bees, and even human social behavior, from which the idea is emerged [14, 7, 22]. PSO is a population-based optimization tool, which could be implemented and applied easily to solve various function optimization problems, or the problems that can be transformed to function optimization problems. As an algorithm, the main strength of PSO is its fast convergence, which compares favorably with many global optimization algorithms like Genetic Algorithms (GA) [13], Simulated Annealing (SA) [20, 27] and other global optimization algorithms. For applying PSO successfully, one of the key issues is finding how to map the problem solution into the PSO particle, which directly affects its feasibility and performance.

Ant Colony Optimization (ACO) deals with artificial systems that is inspired from the foraging behavior of real ants, which are used to solve discrete

A. Abraham et al.: *Swarm Intelligence: Foundations, Perspectives and Applications*, Studies in Computational Intelligence (SCI) **26**, 3–25 (2006)
www.springerlink.com

optimization problems [9]. The main idea is the indirect communication between the ants by means of chemical pheromone trials, which enables them
to find short paths between their nest and food.

This Chapter is organized as follows. Section 1.2 presents the canonical
PSO algorithm and its performance is compared with some global optimization algorithms. Further some extended versions of PSO is presented in Section 1.3 followed by some illustrations/applications in Section 1.4. Section
1.5 presents the ACO algorithm followed by some illustrations/applications
of ACO in Section 1.6 and Section 1.7. Some conclusions are also provided
towards the end, in Section 1.8.

1.2 Canonical Particle Swarm Optimization

1.2.1 Canonical Model

The canonical PSO model consists of a swarm of particles, which are initialized with a population of random candidate solutions. They move iteratively
through the d-dimension problem space to search the new solutions, where the
fitness, f, can be calculated as the certain qualities measure. Each particle has
a position represented by a position-vector $\mathbf{x}_i$ (i is the index of the particle),
and a velocity represented by a velocity-vector $\mathbf{v}_i$. Each particle remembers
its own best position so far in a vector $\mathbf{x}_i^{\#}$, and its j-th dimensional value
is $x_{ij}^{\#}$. The best position-vector among the swarm so far is then stored in a
vector $\mathbf{x}^*$, and its j-th dimensional value is x_j^*. During the iteration time t,
the update of the velocity from the previous velocity to the new velocity is
determined by Eq.(1.1). The new position is then determined by the sum of
the previous position and the new velocity by Eq.(1.2).

$$v_{ij}(t+1) = wv_{ij}(t) + c_1 r_1 (x_{ij}^{\#}(t) - x_{ij}(t)) + c_2 r_2 (x_j^*(t) - x_{ij}(t)). \quad (1.1)$$

$$x_{ij}(t+1) = x_{ij}(t) + v_{ij}(t+1). \quad (1.2)$$

where w is called as the inertia factor, r_1 and r_2 are the random numbers,
which are used to maintain the diversity of the population, and are uniformly
distributed in the interval [0,1] for the j-th dimension of the i-th particle. c_1
is a positive constant, called as coefficient of the self-recognition component,
c_2 is a positive constant, called as coefficient of the social component. From
Eq.(1.1), a particle decides where to move next, considering its own experience,
which is the memory of its best past position, and the experience of its most
successful particle in the swarm. In the particle swarm model, the particle
searches the solutions in the problem space with a range $[-s, s]$ (If the range
is not symmetrical, it can be translated to the corresponding symmetrical
range.) In order to guide the particles effectively in the search space, the
maximum moving distance during one iteration must be clamped in between
the maximum velocity $[-v_{max}, v_{max}]$ given in Eq.(1.3):

$$v_{ij} = sign(v_{ij})min(|v_{ij}|, v_{max}). \tag{1.3}$$

The value of v_{max} is $p \times s$, with $0.1 \le p \le 1.0$ and is usually chosen to be s, i.e. $p = 1$. The pseudo-code for particle swarm optimization algorithm is illustrated in Algorithm 1.

Algorithm 1 Particle Swarm Optimization Algorithm

---

01. Initialize the size of the particle swarm n, and other parameters.
02. Initialize the positions and the velocities for all the particles randomly.
03. While (the end criterion is not met) do
04. $t = t + 1$;
05.   Calculate the fitness value of each particle;
06. $\mathbf{x}^* = argmin_{i=1}^{n}(f(\mathbf{x}^*(t-1)), f(\mathbf{x}_1(t)), f(\mathbf{x}_2(t)), \cdots, f(\mathbf{x}_i(t)), \cdots, f(\mathbf{x}_n(t)))$;
07.   For $i = 1$ to n
08. $\mathbf{x}_i^{\#}(t) = argmin_{i=1}^{n}(f(\mathbf{x}_i^{\#}(t-1)), f(\mathbf{x}_i(t)))$;
09.   For $j = 1$ to $Dimension$
10.   Update the j-th dimension value of $\mathbf{x}_i$ and $\mathbf{v}_i$
10.   according to Eqs.(1.1), (1.2), (1.3);
12.   Next j
13.   Next i
14. End While.

The end criteria are usually one of the following:

- Maximum number of iterations: the optimization process is terminated after a fixed number of iterations, for example, 1000 iterations.
- Number of iterations without improvement: the optimization process is terminated after some fixed number of iterations without any improvement.
- Minimum objective function error: the error between the obtained objective function value and the best fitness value is less than a pre-fixed anticipated threshold.

1.2.2 The Parameters of PSO

The role of inertia weight w, in Eq.(1.1), is considered critical for the convergence behavior of PSO. The inertia weight is employed to control the impact of the previous history of velocities on the current one. Accordingly, the parameter w regulates the trade-off between the global (wide-ranging) and local (nearby) exploration abilities of the swarm. A large inertia weight facilitates global exploration (searching new areas), while a small one tends to facilitate local exploration, i.e. fine-tuning the current search area. A suitable value for the inertia weight w usually provides balance between global and local exploration abilities and consequently results in a reduction of the number of iterations required to locate the optimum solution. Initially, the inertia weight is set as a constant. However, some experiment results indicates that it is better to initially set the inertia to a large value, in order to promote global exploration of the search space, and gradually decrease it to get more refined solutions [11]. Thus, an initial value around 1.2 and gradually reducing towards 0 can be considered as a good choice for w. A better method is to use some adaptive approaches (example: fuzzy controller), in which the parameters can be adaptively fine tuned according to the problems under consideration [24, 16].

The parameters c_1 and c_2, in Eq.(1.1), are not critical for the convergence of PSO. However, proper fine-tuning may result in faster convergence and alleviation of local minima. As default values, usually, $c_1 = c_2 = 2$ are used, but some experiment results indicate that $c_1 = c_2 = 1.49$ might provide even better results. Recent work reports that it might be even better to choose a larger cognitive parameter, c_1, than a social parameter, c_2, but with $c_1 + c_2 \leq 4$ [7].

The particle swarm algorithm can be described generally as a population of vectors whose trajectories oscillate around a region which is defined by each individual's previous best success and the success of some other particle. Various methods have been used to identify some other particle to influence the individual. Eberhart and Kennedy called the two basic methods as "gbest model" and "lbest model" [14]. In the lbest model, particles have information only of their own and their nearest array neighbors' best (lbest), rather than that of the entire group. Namely, in Eq.(1.4), gbest is replaced by lbest in the model. So a new neighborhood relation is defined for the swarm:

$$v_{id}(t+1) = w*v_{id}(t)+c_1*r_1*(p_{id}(t)-x_{id}(t))+c_2*r_2*(p_{ld}(t)-x_{id}(t)). \quad (1.4)$$

$$x_{id}(t+1) = x_{id}(t) + v_{id}(t+1). \quad (1.5)$$

In the gbest model, the trajectory for each particle's search is influenced by the best point found by any member of the entire population. The best particle acts as an attractor, pulling all the particles towards it. Eventually all particles will converge to this position. The lbest model allows each individual to be influenced by some smaller number of adjacent members of the population array. The particles selected to be in one subset of the swarm have no direct relationship to the other particles in the other neighborhood. Typically lbest neighborhoods comprise exactly two neighbors. When the number of neighbors increases to all but itself in the lbest model, the case is equivalent to the gbest model. Some experiment results testified that gbest model converges quickly on problem solutions but has a weakness for becoming trapped in local optima, while lbest model converges slowly on problem solutions but is able to "flow around" local optima, as the individuals explore different regions. The gbest model is recommended strongly for unimodal objective functions, while a variable neighborhood model is recommended for multimodal objective functions.

Kennedy and Mendes [15] studied the various population topologies on the PSO performance. Different concepts for neighborhoods could be envisaged. It can be observed as a spatial neighborhood when it is determined by the Euclidean distance between the positions of two particles, or as a sociometric neighborhood (e.g. the index position in the storing array). The different concepts for neighborhood leads to different neighborhood topologies. Different neighborhood topologies primarily affect the communication abilities and thus the group's performance. Different topologies are illustrated in Fig. 1.1. In the case of a global neighborhood, the structure is a fully connected network where every particle has access to the others' best position (Refer Fig. 1.1(a)). But in local neighborhoods there are more possible variants. In the von Neumann topology (Fig. 1.1(b)), neighbors above, below, and each side on a two dimensional lattice are connected. Fig. 1.1(e) illustrates the von Neumann topology with one section flattened out. In a pyramid topology, three dimensional wire frame triangles are formulated as illustrated in Fig. 1.1(c). As shown in Fig. 1.1(d), one common structure for a local neighborhood is the circle topology where individuals are far away from others (in terms of graph structure, not necessarily distance) and are independent of each other but neighbors are closely connected. Another structure is called wheel (star) topology and has a more hierarchical structure, because all members of the neighborhood are connected to a 'leader' individual as shown in Fig. 1.1(f). So all information has to be communicated though this 'leader', which then compares the performances of all others.

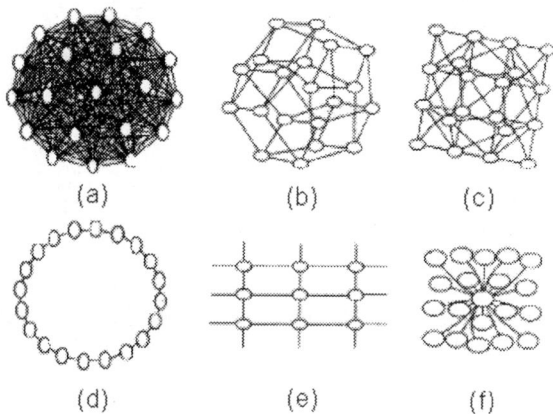

Fig. 1.1. Some neighborhood topologies adapted from [15]

1.2.3 Performance Comparison with Some Global Optimization Algorithms

We compare the performance of PSO with Genetic Algorithm (GA) [6, 13] and Simulated Annealing (SA)[20, 27]. GA and SA are powerful stochastic search and optimization methods, which are also inspired from biological and thermodynamic processes.

Genetic algorithms mimic an evolutionary natural selection process. Generations of solutions are evaluated according to a fitness value and only those candidates with high fitness values are used to create further solutions via crossover and mutation procedures.

Simulated annealing is based on the manner in which liquids freeze or metals re-crystalize in the process of annealing. In an annealing process, a melt, initially at high temperature and disordered, is slowly cooled so that the system at any time is approximately in thermodynamic equilibrium. In terms of computational simulation, a global minimum would correspond to such a "frozen" (steady) ground state at the temperature T=0.

The specific parameter settings for PSO, GA and SA used in the experiments are described in Table 1.1.

Table 1.1. Parameter settings for the algorithms.

Algorithm	Parameter name	Parameter value
GA	Size of the population	20
	Probability of crossover	0.8
	Probability of mutation	0.02
	Scale for mutations	0.1
	Tournament probability	0.7
SA	Number operations before temperature adjustment	20
	Number of cycles	10
	Temperature reduction factor	0.85
	Vector for control step of length adjustment	2
	Initial temperature	50
PSO	Swarm size	20
	Self-recognition coefficient c_1	1.49
	Social coefficient c_2	1.49
	Inertia weight w	$0.9 \rightarrow 0.1$

Benchmark functions:

- Griewank function:
  $$f_1 = \frac{1}{4000} \sum_{i=1}^{n}(x_i)^2 - \prod_{i=1}^{n} cos(\frac{x_i}{\sqrt{i}}) + 1$$
  $\mathbf{x} \in [-300, 300]^n$, $min(f_1(\mathbf{x}^*)) = f_1(\mathbf{0}) = 0$.

- Schwefel 2.26 function:
  $f_2 = 418.9829n - \sum_{i=1}^{n}(x_i sin(\sqrt{|x_i|}))$
  $\mathbf{x} \in [-500, 500]^n$, $min(f_2(\mathbf{x}^*)) = f_2(\mathbf{0}) = 0$.
- Quadric function:
  $f_3 = \sum_{i=1}^{n}(\sum_{j=1}^{i} x_j)^2$
  $\mathbf{x} \in [-100, 100]^n$, $min(f_3(\mathbf{x}^*)) = f_3(\mathbf{0}) = 0$.

Three continuous benchmark functions, i.e. Griewank function, Schwefel 2.26 function and Quadric function, are used to test PSO, GA and SA. Quadric function has a single minimum, while the other two functions are highly multi-modal with multiple local minima. For all the test functions, the goal is to find the global minima. Each algorithm (for each function) was repeated 10 times with different random seeds. Each trial had a fixed number of 18,000 iterations. The objective functions were evaluated 360,000 times in each trial. The swarm size in PSO was 20, population size in GA was 20, the number operations before temperature adjustment in SA was set to 20. Figures 1.2, 1.3 and 1.4 illustrate the mean best function values for the three functions. It is observed that for GA and SA, the solutions get trapped in a local minimum even before 2000 iterations, for high dimensional, multi-modal functions, especially for Schwefel 2.26 function, while PSO performance is much better. For the Quadric function, SA performed well and PSO performance was comparatively poor, as depicted in Fig. 1.4.

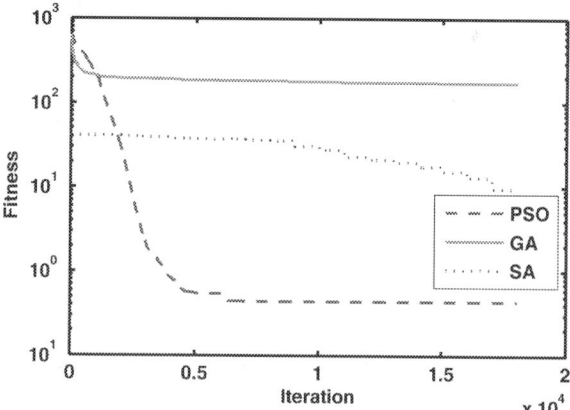

Fig. 1.2. Griewank function performance

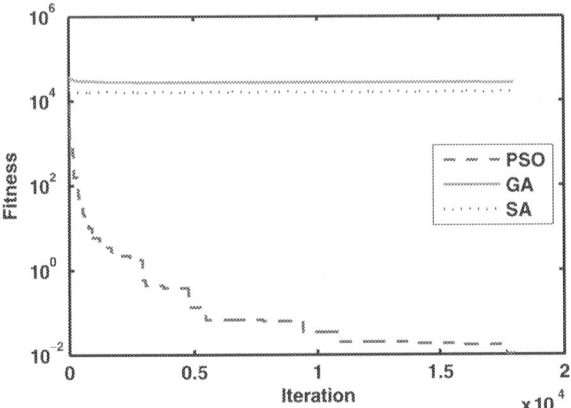

Fig. 1.3. Schwefel function performance

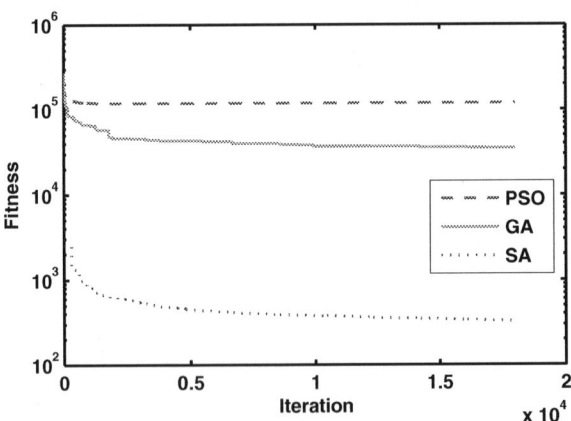

Fig. 1.4. Quadric function performance

1.3 Extended Models of PSO for Discrete Problems

1.3.1 Fuzzy PSO

In the fuzzy PSO model, the representations of the position and velocity of the particles in PSO are extended from real vectors to fuzzy matrices [21]. This is illustrated using the well known job scheduling problem. For a job scheduling problem: the jobs $J = \{J_1, J_2, \cdots, J_n\}$ are to be scheduled on the machines $M = \{M_1, M_2, \cdots, M_m\}$, and the fuzzy scheduling relation from M to J can be expressed as follows:

$$X = \begin{bmatrix} x_{11} & x_{12} & \cdots & x_{1n} \\ x_{21} & x_{22} & \cdots & x_{2n} \\ \vdots & \vdots & \ddots & \vdots \\ x_{m1} & x_{m2} & \cdots & x_{mn} \end{bmatrix}$$

where x_{ij} represents the degree of membership of the i-th element M_i in domain M and the j-th element J_j in domain J to relation X. The fuzzy relation X between M and J has the following meaning: for each element in the matrix X, the element

$$x_{ij} = \mu_R(M_i, J_j), i \in \{1, 2, \cdots, m\}, j \in \{1, 2, \cdots, n\}. \tag{1.6}$$

μ_R is the membership function, the value of x_{ij} means the degree of membership that M_j would process J_i in the feasible schedule solution. The elements of the matrix X should satisfy the following conditions:

$$x_{ij} \in [0, 1], i \in \{1, 2, \cdots, m\}, j \in \{1, 2, \cdots, n\}. \tag{1.7}$$

$$\sum_{i=1}^{m} x_{ij} = 1, i \in \{1, 2, \cdots, m\}, j \in \{1, 2, \cdots, n\}. \tag{1.8}$$

Similarly the velocity of the particle is defined as:

$$V = \begin{bmatrix} v_{11} & v_{12} & \cdots & v_{1n} \\ v_{21} & v_{22} & \cdots & v_{2n} \\ \vdots & \vdots & \ddots & \vdots \\ v_{m1} & v_{m2} & \cdots & v_{mn} \end{bmatrix}$$

The operators of Eqs.(1.1) and (1.2) should be re-defined because the position and velocity have been transformed to the form of matrices. The symbol "$\otimes$" is used to denote the modified multiplication. Let α be a real number, $\alpha \otimes V$ or $\alpha \otimes X$ means all the elements in the matrix V or X are multiplied by α. The symbols "$\oplus$" and "$\ominus$" denote the addition and subtraction between matrices respectively. Suppose A and B are two matrices which denote position or velocity, then $A \oplus B$ and $A \ominus B$ are regular addition and subtraction operation between matrices.

Then we obtain Eqs.(1.9) and (1.10) for updating the positions and velocities of the particles in the fuzzy discrete PSO:

$$V(t+1) = w \otimes V(t) \oplus (c_1 * r_1) \otimes (X^{\#}(t) \ominus X(t)) \oplus (c_2 * r_2) \otimes (X^*(t) \ominus X(t)). \tag{1.9}$$

$$X(t+1) = X(t) \oplus V(t+1)). \tag{1.10}$$

The position matrix may violate the constraints of Eqs.(1.7) and (1.8) after some iterations, so it is necessary to normalize the position matrix. First we make all the negative elements in the matrix become zero. If all elements in a column of the matrix are zero, they need be re-evaluated using a series of random numbers with the interval [0,1]. And then the matrix undergoes the following transformation without violating the constraints:

$$Xnormal = \begin{bmatrix} x_{11}/\sum_{i=1}^{m} x_{i1} & x_{12}/\sum_{i=1}^{m} x_{i2} & \cdots & x_{1n}/\sum_{i=1}^{m} x_{in} \\ x_{21}/\sum_{i=1}^{m} x_{i1} & x_{22}/\sum_{i=1}^{m} x_{i2} & \cdots & x_{2n}/\sum_{i=1}^{m} x_{in} \\ \vdots & \vdots & \ddots & \vdots \\ x_{m1}/\sum_{i=1}^{m} x_{i1} & x_{m2}/\sum_{i=1}^{m} x_{i2} & \cdots & x_{mn}/\sum_{i=1}^{m} x_{in} \end{bmatrix}$$

Since the position matrix indicates the potential scheduling solution, we should "decode" the fuzzy matrix and get the feasible solution. A flag array could be used to record whether we have selected the columns of the matrix and a array to record the solution. First all the columns are not selected, then for each columns of the matrix, we choose the element which has the max value, then mark the column of the max element "selected", and the column number are recorded to the solution array. After all the columns have been processed, we get the feasible solution from the solution array and measure the fitness of the particles.

1.3.2 Binary PSO

The canonical PSO is basically developed for continuous optimization problems. However, lots of practical engineering problems are formulated as combinatorial optimization problems. The binary PSO model was presented by Kennedy and Eberhart, and is based on a very simple modification of the real-valued PSO. Faced with a problem-domain where we cannot fit into some sub-space of the real-valued n-dimensional space, which is required by the PSO, odds are that we can use a binary PSO instead. All we must provide, is a mapping from this given problem-domain to the set of bit strings. As with the canonical PSO, a fitness function f must be defined. In the binary PSO, we can define a particle's position and velocity in terms of changes of probabilities that will be in one state or the other, i.e. yes or no, true or false, or making some other decision. When the particle moves in a state space restricted to zero and one on each dimension, the change of probability with time steps is defined as follows:

$$P(x_{ij}(t+1) = 1) = f(x_{ij}(t), v_{ij}(t), x_{ij}^{\#}(t), x_{j}^{*}(t)). \tag{1.11}$$

where the probability function is usually

$$sign(v_{ij}(t+1) = 1) = \frac{1}{1 + e^{-v_{ij}(t)}}. \tag{1.12}$$

At each time step, each particle updates its velocity and moves to a new position according to Eqs.(1.13) and (1.14):

$$v_{ij}(t+1) = wv_{ij}(t) + c_1 r_1(x_{ij}^{\#}(t) - x_{ij}(t)) + c_2 r_2(x_{j}^{*}(t) - x_{ij}(t)). \tag{1.13}$$

$$x_i(t+1) = \begin{matrix} 1 \text{ if } \rho \leq s(v_i(t)), \\ 0 \text{ otherwise}. \end{matrix} \qquad (1.14)$$

where c_1, c_2 are learning factors; w is inertia factor; r_1, r_2, ρ are random functions in the closed interval $[0, 1]$.

1.4 Applications of Particle Swarm Optimization

1.4.1 Job Scheduling on Computational Grids

Grid computing is a computing framework to meet the growing computational demands. Essential grid services contain more intelligent functions for resource management, security, grid service marketing, collaboration and so on. The load sharing of computational jobs is the major task of computational grids [2].

To formulate our objective, define $C_{i,j}$ $(i \in \{1, 2, \cdots, m\}, j \in \{1, 2, \cdots, n\})$ as the completion time that the grid node G_i finishes the job J_j, $\sum C_i$ represents the time that the grid node G_i finishes all the scheduled jobs. Define $C_{max} = max\{\sum C_i\}$ as the makespan, and $\sum_{i=1}^{m}(\sum C_i)$ as the flowtime. An optimal schedule will be the one that optimizes the flowtime and makespan. The conceptually obvious rule to minimize $\sum_{i=1}^{m}(\sum C_i)$ is to schedule Shortest Job on the Fastest Node (SJFN). The simplest rule to minimize C_{max} is to schedule the Longest Job on the Fastest Node (LJFN). Minimizing $\sum_{i=1}^{m}(\sum C_i)$ asks the average job finishes quickly, at the expense of the largest job taking a long time, whereas minimizing C_{max}, asks that no job takes too long, at the expense of most jobs taking a long time. Minimization of C_{max} will result in the maximization of $\sum_{i=1}^{m}(\sum C_i)$.

To illustrate the performance of the algorithms, we considered a finite number of grid nodes and assumed that the processing speeds of the grid nodes (cput) and the job lengths (processing requirements in cycles) are known. Specific parameter settings of the three considered algorithms (PSO, GA and SA) are described in Table 1.1. The parameters used for the ACO algorithm are as follows:

Number of ants $= 5$
Weight of pheromone trail $\alpha = 1$
Weight of heuristic information $\beta = 5$
Pheromone evaporation parameter $\rho = 0.8$
Constant for pheromone updating $Q = 10$

Each experiment (for each algorithm) was repeated 10 times with different random seeds. Each trial had a fixed number of $50 * m * n$ iterations (m is the number of the grid nodes, n is the number of the jobs). The makespan values of the best solutions throughout the optimization run were recorded. And the averages and the standard deviations were calculated from the 10 different trials. The standard deviation indicates the differences in the results during the 10 different trials. In a grid environment, the main emphasis will be to generate the schedules at a minimal amount of time. So the completion time for 10 trials were used as one of the criteria to improve their performance.

We tested a small scale job scheduling problem involving 3 machines and 13 jobs (3,13) and 5 machines and 100 jobs (5,100). Fig. 1.5 illustrates the performance of the four algorithms for (3,13). The results for 10 GA runs were {47, 46, 47, 47.3333, 46, 47, 47, 47, 47.3333, 49}, with an average value of 47.1167. The results of 10 SA runs were {46.5, 46.5, 46, 46,46, 46.6667, 47, 47.3333, 47, 47}with an average value of 46.6. The results of 10 PSO runs were {46, 46, 46, 46, 46.5, 46.5, 46.5, 46, 46.5, 46.6667}, with an average value of 46.2667. The results of 10 ACO runs were {46, 46, 46, 46, 46.5, 46.5, 46.5, 46, 46, 46.5}, with an average value of 46.2667. The optimal result is supposed to be 46. While GA provided the best results twice, SA, PSO, ACO provided the best results three, five and six times respectively. Empirical results are summarized in Table 1.2 for (3,13) and (5,100).

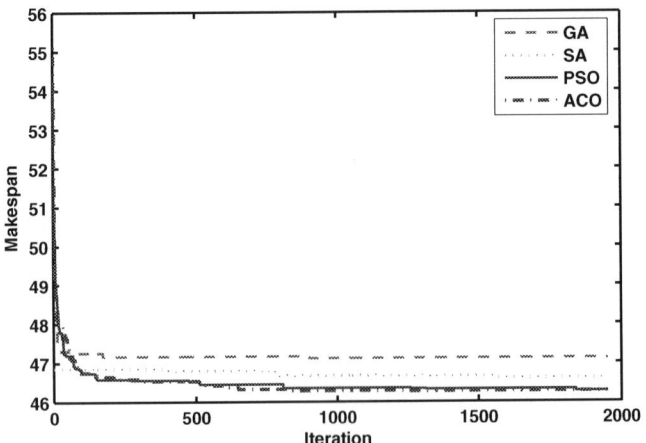

Fig. 1.5. Performance for job scheduling (3,13)

1.4.2 PSO for Data Mining

Data mining and particle swarm optimization may seem that they do not have many properties in common. However, they can be used together to form a method which often leads to the result, even when other methods would be too expensive or difficult to implement. Ujjinn and Bentley [28] provided internet-based recommender system, which employs a particle swarm optimization algorithm to learn personal preferences of users and provide tailored

Table 1.2. Comparing the performance of the considered algorithms.

Algorithm	Item	Instance	
		(3,13)	(5,100)
GA	Average makespan	47.1167	85.7431
	Standard Deviation	±0.7700	±0.6217
	Time	302.9210	2415.9
SA	Average makespan	46.6000	90.7338
	Standard Deviation	±0.4856	±6.3833
	Time	332.5000	6567.8
PSO	Average makespan	46.2667	84.0544
	Standard Deviation	±0.2854	±0.5030
	Time	106.2030	1485.6
ACO	Average makespan	46.2667	88.1575
	Standard Deviation	±0.2854	±0.6423
	Time	340.3750	6758.3

suggestions. Omran et al. [19] used particle swarm to implement image clustering. When compared with K-means, Fuzzy C-means, K-Harmonic means and genetic algorithm approaches, in general, the PSO algorithm produced better results with reference to inter- and intra-cluster distances, while having quantization errors comparable to the other algorithms. Sousa et al. [25] proposed the use of the particle swarm optimizer for data mining. Tested against genetic algorithm and Tree Induction Algorithm (J48), the obtained results indicates that particle swarm optimizer is a suitable and competitive candidate for classification tasks and can be successfully applied to more demanding problem domains. The basic idea of combining particle swarm optimization with data mining is quite simple. To extract this knowledge, a database may be considered as a large search space, and a mining algorithm as a search strategy. PSO makes use of particles moving in an n-dimensional space to search for solutions for an n-variable function (that is fitness function) optimization problem. The datasets are the sample space to search and each attribute is a dimension for the PSO-miner. During the search procedure, each particle is evaluated using the fitness function which is a central instrument in the algorithm. Their values decide the swarm's performance. The fitness function measures the predictive accuracy of the rule for data mining, and it is given by Eq.(1.15):

$$predictive\_accuracy = \frac{|A \wedge C| - 1/2}{|A|} \qquad (1.15)$$

where $|A \wedge C|$ is the number of examples that satisfy both the rule antecedent and the consequent, and $|A|$ is the number of cases that satisfy only the rule antecedent. The term $1/2$ is subtracted in the numerator of Eq.(1.15) to penalize rules covering few training examples. PSO usually search the minimum for the problem space considered. So we use predictive_accuracy to the power minus one as fitness function in PSO-miner.

Rule pruning is a common technique in data mining. The main goal of rule pruning is to remove irrelevant terms that might have been unduly included in the rules. Rule pruning potentially increases the predictive power of the rule, helping to avoid its over-fitting to the training data. Another motivation for rule pruning is that it improves the simplicity of the rule, since a shorter rule is usually easier to be understood by the user than a longer one. As soon as the current particle completes the construction of its rule, the rule pruning procedure is called. The quality of a rule, denoted by Q, is computed by the formula: $Q = sensitivity \cdot specificity$ [17]. Just after the covering algorithm returns a rule set, another post-processing routine is used: rule set cleaning, where rules that will never be applied are removed from the rule set. The purpose of the validation algorithm is to statistically evaluate the accuracy of the rule set obtained by the covering algorithm. This is done using a method known as tenfold cross validation [29]. Rule set accuracy is evaluated and presented as the percentage of instances in the test set correctly classified. In order to classify a new test case, unseen during training, the discovered rules are applied in the order they were discovered.

The performance of PSO-miner was evaluated using four public-domain data sets from the UCI repository [4]. The used parameters' settings are as following: swarm size=30; $c_1 = c_2 = 2$; maximum position=5; maximum velocity=0.1∼0.5; maximum uncovered cases = 10 and maximum number of iterations=4000. The results are reported in Table 1.3. The algorithm is not only simple than many other methods, but also is a good alternative method for data mining.

Table 1.3. Results of PSO-miner

Data set	Predictive accuracy	Number of rules	Number of terms / Number of rules
Wisconsin breast cancer	92.65 ± 0.61	5.70 ± 0.20	1.63
Dermatology	92.65 ± 2.37	7.40 ± 0.19	2.99
Hepatitis	83.65 ± 3.13	3.30 ± 0.15	1.58
Cleveland heart disease	53.50 ± 0.61	9.20 ± 0.25	1.71

1.5 Ant Colony Optimization

In nature, ants usually wander randomly, and upon finding food return to their nest while laying down pheromone trails. If other ants find such a path (pheromone trail), they are likely not to keep travelling at random, but to instead follow the trail, returning and reinforcing it if they eventually find food. However, as time passes, the pheromone starts to evaporate. The more time it takes for an ant to travel down the path and back again, the more time the pheromone has to evaporate (and the path to become less prominent). A shorter path, in comparison will be visited by more ants (can be described as a loop of positive feedback) and thus the pheromone density remains high for a longer time.

ACO is implemented as a team of intelligent agents which simulate the ants behavior, walking around the graph representing the problem to solve using mechanisms of cooperation and adaptation. ACO algorithm requires to define the following [5, 10]:

- The problem needs to be represented appropriately, which would allow the ants to incrementally update the solutions through the use of a probabilistic transition rules, based on the amount of pheromone in the trail and other problem specific knowledge. It is also important to enforce a strategy to construct only valid solutions corresponding to the problem definition.
- A problem-dependent heuristic function η that measures the quality of components that can be added to the current partial solution.
- A rule set for pheromone updating, which specifies how to modify the pheromone value τ.
- A probabilistic transition rule based on the value of the heuristic function η and the pheromone value τ that is used to iteratively construct a solution.

ACO was first introduced using the Travelling Salesman Problem (TSP). Starting from its start node, an ant iteratively moves from one node to another. When being at a node, an ant chooses to go to a unvisited node at time t with a probability given by

$$p_{i,j}^k(t) = \frac{[\tau_{i,j}(t)]^\alpha [\eta_{i,j}(t)]^\beta}{\sum_{l \in N_i^k} [\tau_{i,j}(t)]^\alpha [\eta_{i,j}(t)]^\beta} \qquad j \in N_i^k \tag{1.16}$$

where N_i^k is the feasible neighborhood of the ant_k, that is, the set of cities which ant_k has not yet visited; $\tau_{i,j}(t)$ is the pheromone value on the edge (i, j) at the time t, α is the weight of pheromone; $\eta_{i,j}(t)$ is a priori available heuristic information on the edge (i, j) at the time t, β is the weight of heuristic information. Two parameters α and β determine the relative influence of pheromone trail and heuristic information. $\tau_{i,j}(t)$ is determined by

$$\tau_{i,j}(t) = \rho\tau_{i,j}(t-1) + \sum_{k=1}^{n} \Delta\tau_{i,j}^k(t) \qquad \forall(i, j) \tag{1.17}$$

$$\Delta\tau_{i,j}^k(t) = \begin{array}{ll} \frac{Q}{L_k(t)} & \text{if the edge } (i, j) \text{ chosen by the } ant_k \\ 0 & \text{otherwise} \end{array} \tag{1.18}$$

where ρ is the pheromone trail evaporation rate $(0 < \rho < 1)$, n is the number of ants, Q is a constant for pheromone updating.

More recent work has seen the application of ACO to other problems [12, 26]. A generalized version of the pseudo-code for the ACO algorithm with reference to the TSP is illustrated in Algorithm 2.

Algorithm 2 Ant Colony Optimization Algorithm

---
01. Initialize the number of ants n, and other parameters.
02. While (the end criterion is not met) do
03. $t = t + 1$;
04.   For $k= 1$ to n
05. ant_k is positioned on a starting node;
06.   For $m= 2$ to $problem\_size$
07.   Choose the state to move into
08.   according to the probabilistic transition rules;
09.   Append the chosen move into $tabu_k(t)$ for the ant_k;
10.   Next m
11.   Compute the length $L_k(t)$ of the tour $T_k(t)$ chosen by the ant_k;
12.   Compute $\Delta\tau_{i,j}(t)$ for every edge (i,j) in $T_k(t)$ according to Eq.(1.18);
13.   Next k
14.   Update the trail pheromone intensity for every edge (i,j) according to Eq.(1.17);
15.   Compare and update the best solution;
16. End While.

1.6 Ant Colony Algorithms for Optimization Problems

1.6.1 Travelling Salesman Problem (TSP)

Given a collection of cities and the cost of travel between each pair of them, the travelling salesman problem is to find the cheapest way of visiting all of the cities and returning to the starting point. It is assumed that the travel costs are symmetric in the sense that travelling from city X to city Y costs just as much as travelling from Y to X. The parameter settings used for ACO algorithm are as follows:

Number of ants = 5
Maximum number of iterations = 1000
$\alpha = 2$
$\beta = 2$
$\rho = 0.9$
$Q = 10$

A TSP with 20 cities (Table 1.4) is used to illustrate the ACO algorithm.

The best route obtained is depicted as $1 \to 14 \to 11 \to 4 \to 8 \to 10 \to 15 \to$ $19 \to 7 \to 18 \to 16 \to 5 \to 13 \to 20 \to 6 \to 17 \to 9 \to 2 \to 12 \to 3 \to 1$, and is illustrated in Fig. 1.6 with a cost of 24.5222. The search result for a TSP for 198 cities is illustrated in Figure 1.7 with a total cost of 19961.3045.

Table 1.4. A TSP (20 cities)

Cities 1	2	3	4	5	6	7	8	9	10
x	5.2940	4.2860	4.7190	4.1850	0.9150	4.7710	1.5240	3.4470	3.7180	2.6490
y	1.5580	3.6220	2.7740	2.2300	3.8210	6.0410	2.8710	2.1110	3.6650	2.5560

Cities 11	12	13	14	15	16	17	18	19	20
x	4.3990	4.6600	1.2320	5.0360	2.7100	1.0720	5.8550	0.1940	1.7620	2.6820
y	1.1940	2.9490	6.4400	0.2440	3.1400	3.4540	6.2030	1.8620	2.6930	6.0970

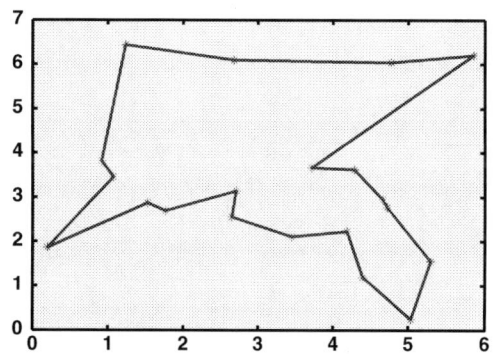

Fig. 1.6. An ACO solution for the TSP (20 cities)

1.6.2 Quadratic Assignment Problem (QAP)

Quadratic assignment problems model many applications in diverse areas such as operations research, parallel and distributed computing, and combinatorial data analysis. There are a set of n facilities and a set of n locations. For each pair of locations a distance is specified and for each pair of facilities a weight or flow is specified (e.g., the amount of supplies transported between the two facilities). The problem is to assign all facilities to different locations with the goal of minimizing the sum of the distances multiplied by the corresponding

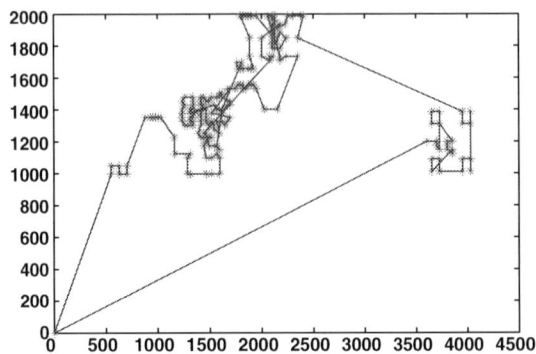

Fig. 1.7. An ACO solution for the TSP (198 cities)

flows. A QAP is used to demonstrate the validity of ACO and its distance/flow matrix for $9 * 9$ assignment is illustrated below:

$$
\begin{array}{c c c c c c c c c c}
 & 1 & 2 & 3 & 4 & 5 & 6 & 7 & 8 & 9 \\
1 & * & 1 & 2 & 3 & 1 & 2 & 3 & 4 & 5 \\
2 & 5 & * & 1 & 2 & 2 & 1 & 2 & 3 & 4 \\
3 & 2 & 3 & * & 1 & 3 & 2 & 1 & 2 & 3 \\
4 & 4 & 0 & 0 & * & 4 & 3 & 2 & 1 & 2 \\
5 & 1 & 2 & 0 & 5 & * & 1 & 2 & 3 & 2 \\
6 & 0 & 2 & 0 & 2 & 10 & * & 1 & 2 & 1 \\
7 & 0 & 2 & 0 & 2 & 0 & 5 & * & 1 & 2 \\
8 & 6 & 0 & 5 & 10 & 0 & 1 & 10 & * & 1 \\
9 & 0 & 4 & 0 & 2 & 5 & 0 & 3 & 8 & *
\end{array}
$$

The parameter settings used for ACO algorithm are as follows:

Number of ants = 5
Maximum number of iterations = 100,000
$\alpha = 1$
$\beta = 5$
$\rho = 0.8$
$Q = 10$

Using ACO, the cheapest cost obtained = 144 and iteration time = 22445. Assignment results are depicted below:

Dept 1 → Site 5; Dept 2 → Site 2; Dept 3 → Site 1; Dept 4 → Site 9; Dept 5 → Site 4; Dept 6 → Site 8; Dept 7 → Site 7; Dept 8 → Site 6; Dept 9 → Site 3.

1.7 Ant Colony Algorithms for Data Mining

The study of ant colonies behavior and their self-organizing capabilities is of interest to knowledge retrieval/management and decision support systems sciences, because it provides models of distributed adaptive organization, which are useful to solve difficult classification, clustering and distributed control problems.

Ant colony based clustering algorithms have been first introduced by Deneubourg et al. [8] by mimicking different types of naturally-occurring emergent phenomena. Ants gather items to form heaps (clustering of dead corpses or cemeteries) observed in the species of *Pheidole Pallidula* and *Lasius Niger*. If sufficiently large parts of corpses are randomly distributed in space, the workers form cemetery clusters within a few hours, following a behavior similar to segregation. If the experimental arena is not sufficiently large, or if it contains spatial heterogeneities, the clusters will be formed along the edges of the arena or, more generally, following the heterogeneities. The basic mechanism underlying this type of aggregation phenomenon is an attraction between dead items mediated by the ant workers: small clusters of items grow by attracting workers to deposit more items. It is this positive and auto-catalytic feedback that leads to the formation of larger and larger clusters.

A sorting approach could be also formulated by mimicking ants that discriminate between different kinds of items and spatially arrange them according to their properties. This is observed in the *Leptothorax unifasciatus* species where larvae are arranged according to their size.

The general idea for data clustering is that isolated items should be picked up and dropped at some other location where more items of that type are present. Ramos et al. [23] proposed *ACLUSTER* algorithm to follow real ant-like behaviors as much as possible. In that sense, bio-inspired spatial transition probabilities are incorporated into the system, avoiding randomly moving agents, which encourage the distributed algorithm to explore regions manifestly without interest. The strategy allows guiding ants to find clusters of objects in an adaptive way.

In order to model the behavior of ants associated with different tasks (dropping and picking up objects), the use of combinations of different response thresholds was proposed. There are two major factors that should influence any local action taken by the ant-like agent: the number of objects in its neighborhood, and their similarity. Lumer and Faieta [18] used an average similarity, mixing distances between objects with their number, incorporating it simultaneously into a response threshold function like the algorithm proposed by Deneubourg et al. [8]. *ACLUSTER* [23] uses combinations of two independent response threshold functions, each associated with a different environmental factor depending on the number of objects in the area, and their similarity. Reader may consult [23] for the technical details of *ACLUSTER*.

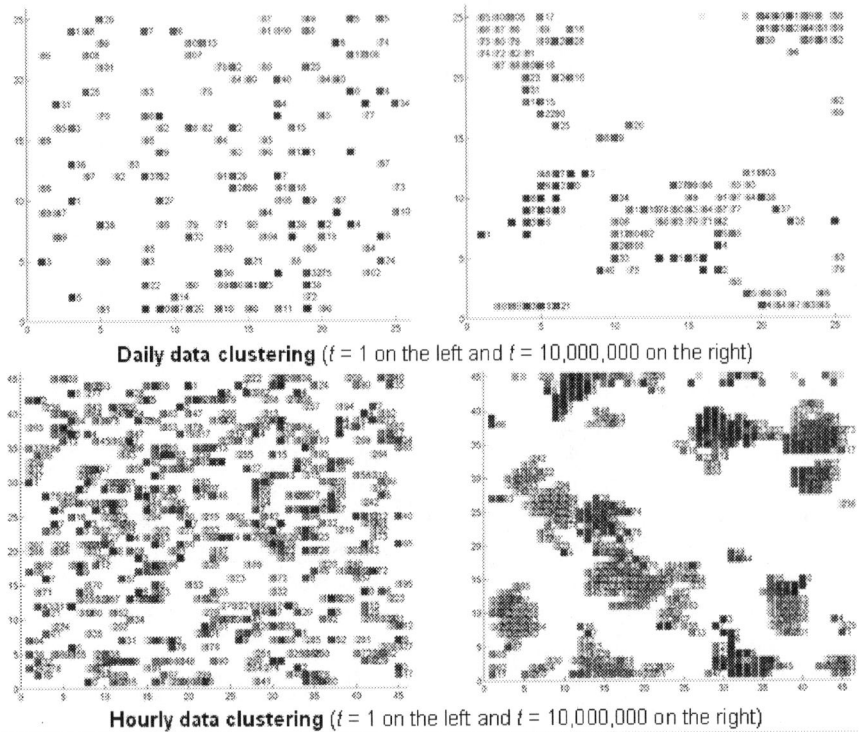

Daily data clustering ($t = 1$ on the left and $t = 10,000,000$ on the right)

Hourly data clustering ($t = 1$ on the left and $t = 10,000,000$ on the right)

Fig. 1.8. Clustering of Web server visitors using ant colony algorithm (adapted from [3])

1.7.1 Web Usage Mining

Web usage mining has become very critical for effective Web site management, creating adaptive Web sites, business and support services, personalization, network traffic flow analysis etc. [3]. Accurate Web usage information could help to attract new customers, retain current customers, improve cross marketing/sales, effectiveness of promotional campaigns, track leaving customers and find the most effective logical structure for their Web space. User profiles could be built by combining users' navigation paths with other data features, such as page viewing time, hyperlink structure, and page content.

Abraham and Ramos [3] used an ant colony clustering algorithm to discover Web usage patterns (data clusters). The task is to cluster similar visitors accessing the web server based on geographical location, type of information requested, time of access and so on. Web log data of a University server from January 01, 2002 to July 0, 2002 was used in the experiments. The log data was categorized into daily and hourly and for each data set the *ACLUSTER* was run twice for 10,00,000 iterations. A 2D classification space is used which is non-parametric and toroidal.

Experiment results for the daily and hourly Web traffic data are illustrated in Fig. 1.8. Fig. 1.8, at the top, represent the spatial distribution of daily Web traffic data on a 25×25 non-parametric toroidal grid. At $t=1$, data items are randomly allocated and 14 ants were deployed and as time evolved, several homogenous clusters emerged. Figure 1.8, at the bottom, represent the spatial distribution of hourly Web traffic data on a 45×45 non-parametric toroidal grid. At $t=1$, data items are randomly allocated and 48 ants were deployed and as time evolved, several homogenous clusters emerged. Reader may consult [3] for detailed results of the different clustering methods.

Clustering results clearly show that ant colony clustering performs well when compared to other clustering methods namely self-organizing maps and evolutionary-fuzzy clustering approach [1].

1.8 Summary

This chapter introduced the theoretical foundations of swarm intelligence with a focus on the implementation and illustration of particle swarm optimization and ant colony optimization algorithms. We provided the design and implementation methods for some applications involving function optimization problems, real world applications and data mining. Results were analyzed, discussed and their potentials were illustrated.

Acknowledgements

First author was supported by the International Joint Research Grant of the IITA (Institute of Information Technology Assessment) foreign professor invitation program of the MIC (Ministry of Information and Communication), South Korea.

References

1. Abraham A (2003) Business intelligence from web usage mining, Journal of Information and Knowledge Management (JIKM), World Scientific Publishing Co., Singapore, 2(4)375-390.
2. Abraham A, Buyya R and Nath B (2000) Nature's heuristics for scheduling jobs on computational grids. Proceedings of the 8th IEEE International Conference on Advanced Computing and Communications, 45-52.
3. Abraham A and Ramos V (2003) Web usage mining using artificial ant colony clustering and genetic programming. Proceesings of IEEE Congress on Evolutionary Computation, Australia, 1384-1391.
4. Blake C, Keogh E and Merz C J (2003) UCI repository of machine learning databases. http://ww.ic.uci.edu/ mlearn/MLRepository.htm.

5. Bonabeau E, Dorigo M and Theraulaz G (1999) Swarm Intelligence: From Natural to Artificial Systems. New York, NY: Oxford University Press.
6. Cantu-Paz E (2000) Efficient and Accurate Parallel Genetic Algorithms. Kluwer Academic publishers.
7. Clerc M and Kennedy J (2002) The particle swarm-explosion, stability, and convergence in a multidimensional complex space. IEEE Transactions on Evolutionary Computation, 6(1):58-73.
8. Deneubourg J-L, Goss S, Franks N, at el. (1991) The dynamics of collective sorting: Robot-like ants and ant-like robots. Proceedings of the First International Conference on Simulation of Adaptive Behaviour: From Animals to Animats, Cambridge, MA: MIT Press, 1, 356-365.
9. Dorigo M, Maniezzo V and Colorni A (1996). Ant system: optimization by a colony of cooperating agents. IEEE Transactions on Systems, Man, and Cybernetics-Part B, 26(1):29-41.
10. Dorigo M and Stützle T (2004), Ant Colony Optimization, MIT Press, 2004.
11. Eberhart R C and Shi Y (2002) Comparing inertia weights and constriction factors in particle swarm optimization. Proceedings of IEEE International Congress on Evolutionary Computation, 84-88.
12. Gambardella L M and Dorigo M (1995) Ant-Q: A reinforcement learning approach to the traveling salesman problem. Proceedings of the 11th International Conference on Machine Learning, 252-260.
13. Goldberg D E (1989) Genetic Algorithms in search, optimization, and machine learning. Addison-Wesley Publishing Corporation, Inc.
14. Kennedy J and Eberhart R (2001) Swarm intelligence. Morgan Kaufmann Publishers, Inc., San Francisco, CA.
15. Kennedy J and Mendes R (2002) Population structure and particle swarm performance. Proceeding of IEEE conference on Evolutionary Computation, 1671-1676.
16. Liu H and Abraham A (2005) Fuzzy Turbulent Particle Swarm Optimization. Proceeding of the 5th International Conference on Hybrid Intelligent Systems, Brazil, IEEE CS Press, USA.
17. Lopes H S, Coutinho M S and Lima W C (1998) An evolutionary approach to simulate cognitive feedback learning in medical domain. Genetic Algorithms and Fuzzy Logic Systems: Soft Computing Perspectives, World Scientific, 193-207.
18. Lumer E D and Faieta B (1994) Diversity and Adaptation in Populations of Clustering Ants. Cli D, Husbands P, Meyer J and Wilson S (Eds.), Proceedings of the Third International Conference on Simulation of Adaptive Behaviour: From Animals to Animats 3, Cambridge, MA: MIT Press, 501-508.
19. Omran M, Engelbrecht P A and Salman A (2005) Particle swarm optimization for image clustering. International Journal of Pattern Recognition and Artificial Intelligence, 19(3):297-321.
20. Orosz J E and Jacobson S H (2002) Analysis of static simulated annealing algorithms. Journal of Optimzation theory and Applications, 115(1):165-182.
21. Pang W, Wang K P, Zhou C G, at el. (2004) Fuzzy discrete particle swarm optimization for solving traveling salesman problem. Proceedings of the 4th International Conference on Computer and Information Technology, IEEE CS Press.
22. Parsopoulos K E and Vrahatis M N (2004) On the computation of all global minimizers through particle swarm optimization. IEEE Transactions on Evolutionary Computation, 8(3):211-224.

23. Ramos V, Muge F, Pina P (2002) Self-organized data and image retrieval as a consequence of inter-dynamic synergistic relationships in artificial ant colonies. Soft Computing Systems - Design, Management and Applications, Proceedings of the 2nd International Conference on Hybrid Intelligent Systems, IOS Press, 500-509.
24. Shi Y H and Eberhart R C (2001) Fuzzy adaptive particle swarm optimization. Proceedings of IEEE International Conference on Evolutionary Computation, 101-106.
25. Sousa T, Silva A, Neves A (2004) Particle swarm based data mining algorithms for classification tasks. Parallel Computing, 30:767-783.
26. Stützle T and Hoo H H (2000) MAX-MIN ant system. Future Generation Computer Systems, 16:889-914.
27. Triki E, Collette Y and Siarry P (2005) A theoretical study on the behavior of simulated annealing leading to a new cooling schedule. European Journal of Operational Research, 166:77-92.
28. Ujjin S and Bentley J P (2003) Particle swarm optimization recommender system. Proceeding of IEEE International conference on Evolutionary Computation, 124-131.
29. Witten Ian H and Frank E (1999) Data mining - Practical Machine Learning Tools and Techniques with Java Implementations. CA: Morgan Kauffmann.

2

Waves of Swarm Particles (WoSP)

Tim Hendtlass

Faculty of Information and Communication Technologies,
Swinburne University of Technology, Hawthorn, Australia.
thendtlass@swin.edu.au

The conventional particle swarm optimisation algorithm has proved very sucessful at finding a good optimum in problem spaces of low to medium complexity. However problem spaces with many optima can prove difficult, especially if the dimensionality of the problem space is high. The probability that the conventional particle swarm algorithm will converge to a sub-optimal position is unacceptably high. In this chapter an adaption of the conventional particle swarm algorithm is introduced that converts the behaviour from the conventional search and converge to an endless cycle of search, converge and then diverge to carry on searching. After introducing this new waves of swarm particles (WoSP) algorithm, its behaviour on a number of problem spaces is presented. The simpler of these problem spaces have been chosen to explore the parameters of the new algorithm, but the last problem spaces have been chosen to show the remarkable performance of the algorithm on highly deceptive multi dimensional problem spaces with extreme numbers of local optima.

2.1 The Conventional Particle Swarm Algorithm

Many algorithms are the result of biological inspiration and particle swarm optimisation (PSO) is no exception. However, the PSO algorithm has slightly different end goals to the biological behaviour that provides its inspiration and so needs to differ from nature in some, perhaps un-biological, ways. PSO takes its inspiration from the flocking of birds and fish, which in the real world flock for the purposes of protection and efficient searching for food. In the real world, the swarm needs to be compact for protection; once food is found the flock should settle to feed. In artificial particle swarm optimisation the algorithm seeks to find an optimum position, rather than the protection or food sought in the natural environment. For PSO the correct behaviour once an optimum is found is not for all the particles in the swarm to converge

T. Hendtlass: *Waves of Swarm Particles (WoSP)*, Studies in Computational Intelligence (SCI)
26, 27–58 (2006)
www.springerlink.com

on this, possibly local, optimum as the goal is to check many optima in the hope of finding the global optimum. Instead of converging, once an optimum has been found, the particles should immediately disperse to look for another, perhaps better, optimum.

If we model our algorithm too closely on the behaviour of birds and fish we run the risk that we will achieve those aspects of the natural behaviour that we don't want at the expense of the artificial behaviour that we do want. While retaining the components that give the natural swarm its efficient search capability, we should add such un-biological components as necessary so as to modify the natural behaviour into the type of behaviour we desire. This chapter describes adding two such additional components and the modified behaviour (compared to the biological) that we achieve by adding them. It will be seen that this new behaviour is particularly suited to searching through a problem space with multiple optima.

In the particle swarm algorithm the position of a particular particle in some way encodes a possible solution to the problem for which we seek, ideally an optimal, but at least a very good solution. Particles move under the influence of a number of factors in such a way that they tend to converge to an optimum. A number of variants of the particle swarm algorithm have been proposed since the original algorithm was introduced by [6]. All of these try to balance several aspects of the behaviour. See, for example, [2, 7, 3]

Firstly all particles are assumed to have momentum so that they cannot change direction instantaneously and each time their velocity is updated it must contain a component that is in the same direction as their previously calculated velocity. This effectively provides a low pass filter to any change in their velocity and smooths the track the particle follows through the problem space.

Secondly, all particles try to exploit at least one good position already found by some particle in the swarm. Often the position that is exploited is the best position yet found by any member of the swarm. In this case all the swarm members know the currently best position found by any swarm member and are attracted to (will have a component toward) this position. This obviously requires communication between the members of the swarm and some sort of collective memory as to the current global best (*gbest*) position. Alternatively each particle can experience an attraction back to the best place yet found by this particle. This uses a set of personal best (*Pbest*) positions, one for each particle and on its own would result them exploring independently without any input for the other swarm members. When combined with the momentum it can produce particles that explore around the vicinity of the best place they themselves have yet found. Another option is discussed below.

With only the use of momentum and *gbest*, particles would engage in a headlong rush toward the first reasonable position found, only changing this if some particle happens upon a better position during this rush. Eventually all particles would reach the same best-known position and exploration would stop. The particles would in time come to rest as, once a particle over shot a position, it would be attracted back to it and, with the momentum term being less than one, the velocity would drop with each reversal. This behaviour would mimic real life birds settling at the best-known food source.

Particles are usually started at random positions and the use of *pbest* rather than *gbest* can result in each particle finally settling on the best position that it itself has discovered. This would give a parallel search and allow a choice to be made from the optima found. However some, if not many, particles may effectively waste their time exploring in regions of poor fitness and the local optimum they find is unlikely to be of real interest. What is needed is some attraction, if not to the absolutely best position known, at least toward a position close to this particle where the fitness is known and better than the fitness it is currently experiencing.

One way to achieve this is to define a local neighbourhood around each particle and for every particle in a neighbourhood to share its fitness with all other particles in its neighbourhood. Particles experiemce an attraction to the best performing particle in its neighbourhood *lbest*. Using all the particles within an actual physical distance would result in the neighbourhood being defined as consisting of this particle and its closest N others. This approach has the advantage of being intuitively obvious but in practice choosing a suitable value for this distance is not always easy. Alternatively, assuming that the particles are identified by some index, the neighbourhood could be defined as consisting of this particle and its all particles whose index is with $N/2$ of this particle's index. The problem with neighbourhoods is that they need to be calculated frequently and so the computational cost of this has to be considered. One way of calculating a unique neighbourhood point for each particle to be attracted toward that uses the performance of all particles but is of modest computational cost is given in [4]. Here the position is found as a centre of attraction, where each particle contributes an amount that is proportional to its fitness and inversely proportional to the distance it is away.

However the neighbourhood is defined, this extra attraction toward a place of at least equal and probably better fitness has the effect of causing local exploration around this point. As, in general, no two particles have the same neighbourhoods and neighbourhoods vary with time, this has the result of some exploitation of the good findings of others but without such a strong drive of all particles to a common point as when using *gbest*. Commonly both *gbest* and *lbest* are used together, with some weighting factor being introduced to set the relative influences of each.

Other variations to a particle's velocity are, of course, possible. For example, the addition of a simple random velocity (of modest magnitude) has the advantage of perturbing a particle's path so that it explores around the direct path it is taking, which may result in the discovery of a good region that would otherwise have been missed. However, unlike the use of *lbest* described above, there is no bias toward better fitness in the random changes made as they have an equal probability of causing a movement toward a region of worse fitness as toward a region of better fitness. On the grounds of computational efficiency the use of a random addition may not be such a good idea as *lbest*. However, adding an element of randomness effectively adds noise to the search process, which has been found to be advantageous in evolution [9]. There are other ways to add randomness, such as using random relative weightings for *gbest* and *lbest* as will be used below.

The continual movement bias toward a position of better fitness in the local based approach is very reminiscent of the way that evolution biases selection to individuals with better fitness. The power of evolution is well known so it should come as no surprise that the use of local attraction also dramatically improves the average fitness of the positions explored. Of course, just like evolution, this may result in exploration stopping at a local optimum, but with a number of different local neighbourhoods in use there is a very good probability that the whole swarm will not get so trapped and that any trapped particle will escape, especially if a *gbest* attraction force is also simultaneously in use.

Equation 2.1 shows one possible update formula that uses momentum as well as an attraction to *gbest* and *lbest* In this equation the velocity $\bar{V}_{T+t}$ of a particle at time $T + t$ is derived from its position $\overline{C_T}$ and velocity $\overline{V_T}$ at time T.

$$V_{T+t} = \chi \left(M \cdot V_t + \frac{rand \cdot B \left(\overline{B} - \overline{C_T}\right)}{t} + \frac{rand \cdot L \left(\overline{L_T} - \overline{C_T}\right)}{t} \right) \quad (2.1)$$

In equation 2.1, $\overline{B}$ is a vector to the best position (*gbest*) found by the swarm so far and $\overline{L_T}$ is a vector to the best position in the local neighbourhood of the particle at time T *(lbest)*. M ($0 \leq M \leq 1$) sets the momentum of the particle and B and L are parameters that set the relative importance of the attraction to the *gbest* and *lbest* positions respectively although these values are moderated by the random numbers *rand* ($0 \leq rand \leq 1$). The time interval t between velocity updates is often taken to be unity and omitted: it is shown explicitly here as the equation is dimensionally inconsistent without it and because it will be of great importance later in this chapter. The constriction factor χ is not defined here, as it will not be used further. A discussion of it will be found in [3].

The use of the two random numbers introduces randomness not as an offset but as a weighting to the influence of *gbest* and *lbest*. These influences are not constant but vary around one half of B and L respectively, unlike momentum that is always uniformly applied.

Note that the magnitude of the attraction toward $\overline{B}$ and $\overline{L_T}$ is a function of how far away these points are from the particle (albeit moderated by a weighting factor and a random number). Thus particle far from, say, the best position $\overline{B}$ will experience a large attraction toward $\overline{B}$, an attraction that will diminish as the particle approaches $\overline{B}$. As a consequence of this strong attraction, the momentum of the particle as it arrives at $\overline{B}$ may be very large and as a result the particle may overfly $\overline{B}$ by a significant distance. The continual attraction toward $\overline{B}$ will finally bring the particle to a stop and then attract it back. Eventually after as series of overshoots and returns the particle will come (almost) to a stop at $\overline{B}$, always assuming a better global position was not found by some swarm member during the time that the particle was traveling to and fro.

Unlike real swarms, where velocities continually change, velocity updates in PSO occur at regular intervals, each separated from the next by a discrete time interval t. During these intervals the velocity is assumed to be unchanging. Thus, when the velocity of the particle is high, the distance traveled between velocity updates will be large. As fitness evaluations are only conducted at these times of velocity update, it is quite possible that the particle could overfly some better position but be unaware of the fact as the transit did not coincide with a fitness evaluation. To minimise the chance of this occurring an upper bound to the velocity a particle might be introduced, and / or a constriction factor χ might be applied that steadily reduces the velocity of particles as time passes.

The classical particle swarm algorithm consists of using equation 1 to track a number of particles as they travel through a multi-dimensional space (called feature or problem space) for which a mapping must exist between any position in this space and a solution to the problem we are trying to optimise. The dimensionality of the feature space need not be (but often is) the same as the number of variables in the problem we wish to solve. All that is required is that the fitness of a particle can be derived by mapping its position to a particular solution and then evaluating that solution. A pseudo code version of a generic classical PSO algorithm is as follows.

Algorithm 3 A generic classical PSO algorithm

1. Position all the particles that make up the swarm randomly through feature space and evaluate the fitness of each particle. Assign the position in feature space that corresponds to the best fitness as *gbest*. Assign random velocities to each particle.
2. Assuming that these velocities remain unchanged though the time interval t, calculate the new position of each particle at the end of this period.
3. Evaluate the fitness of each particle, updating *gbest* if necessary.
4. If the best performance is adequate or enough time has passed without any change to *gbest* exit, else go on to step 5.
5. For each particle in turn, calculate *lbest* and then use equation 1 to calculate the new velocity.
6. Return to step 2.

This algorithm has proved to be highly successful at finding the optima of a range of problems in which one clear optimum exists, but less successfully for problems with a larger number of optima. Less successful in the sense that, while it will always find an optimum, the probability that it will find the global optimum decreases as the number of optima to be explored increases, especially if a significant number of the optima have comparable fitness. For this latter type of problem the PSO algorithm needs to be modified in one of two ways.

The first possibility is to inhibit any tendency of the swarm to converge on one position until sufficient exploration has been undertaken so that one has a reasonably expectation that the point on which the swarm is converging is a very good (hopefully optimal) one. Sharply reducing, if not eliminating, the influence of *gbest* encourages such exploration, but at the very best the number of members in the swarm limits the number of optima that will be explored. Increasing the size of the swarm increases the computational load, particularly when calculating the fitness of a position is computationally expensive.

The second possibility is to stop attempting to find the final answer in a single convergence, but to change the behaviour of the swarm to a series of explore, converge and record cycles, with later cycles involving as little re-exploration of previously discovered optima as possible. The final answer can then be chosen by some external agent from the sequence of optima explored. To achieve this behavioural change, we need to add an attraction that only becomes significant when a swarm has settled on an optimum and which has the effect of dispersing the swarm members to search further. This differs from just reinitialising the swarm and running the algorithm over and over, as the particles will carry information about previous optima with them. The WoSP algorithm adds this extra attraction to the conventional PSO.

2.2 The WoSP Algorithm

There are two extra parts to the WoSP (Waves of Swarm particles) algorithm compared to the conventional swarm algorithm; the first adds a short range force as an extra term to the update equation presented as equation 1 (and a few other detail changes). This has the effect of some particles being vigorously ejected from their current position, especially when the swarm is settling on an optimum. On its own, this is of only occasional use as long as the ejected particle retains knowledge of the optimum to which it was being attracted before the ejection occurred. So additional changes have to be made so that ejected particles get a fair chance to explore. These changes will now be considered in detail.

2.2.1 Adding a Short-Range Force

A short-range force of attraction between particles will only alter the behaviour of particles (as compared to the conventional PSO algorithm) when they are close together. While the probability that this could happen as two particles happen to pass close to each other during their voyages through feature space is not (quite) zero, it is far most likely to occur as the particles converge when the swarm is settling on an optimum. One possibility is to introduce a gravitational style attraction that produces a force of attraction of particle i toward particle j whose magnitude is inversely proportional to some power p of the distance between them. This short-range force (SRF) will produce a velocity component v_{ij} of one particle toward the other that can be represented by:

$$v_{ij} = \frac{SRF_{factor}}{d_{ij}^{SRF_{power}}} \tag{2.2}$$

where d_{ij} is the distance between the particles i and j, and SRF_{factor} is a constant that includes, amongst other things, the assumed mass of the particles.

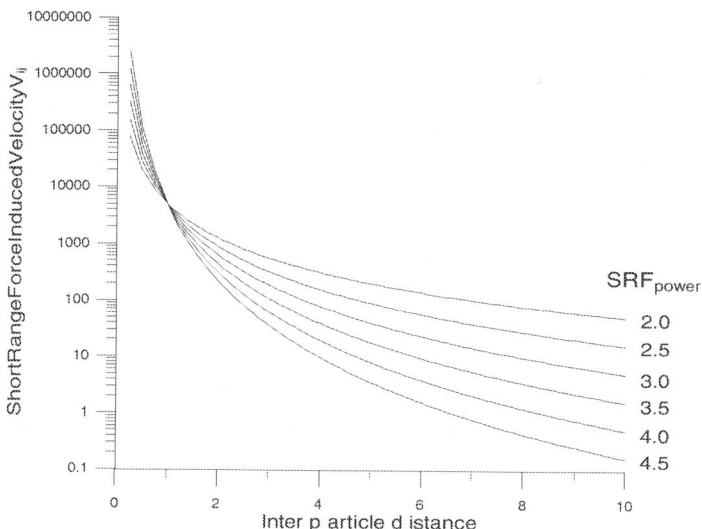

Fig. 2.1. The variation of v_{ij} in equation 2.2 for various values of SRF_{power}

Fig. 2.1 shows the variation of v_{ij} for various values of SRF_{power}. Note that the magnitude of the induced velocity can change by several orders of magnitude for a small variation in the inter particle separation.

This force on its own would cause the particles to tend to enter limit cycles around each other, but when combined with the non-continuous evaluation inherent in all swarm algorithms, the effect is very different.

2.2.2 The Effect of Discrete Evaluation

The short-range force just introduced will have little effect while the swarm is dispersed for either continuous[1] or discrete evaluation. However, as particles approach each other the magnitude of the short-range force will increase significantly, producing a substantial increase in the velocity of the particles toward each other. For discrete evaluation, *by the time of the next evaluation*, particles may have passed each other and be at such a distance apart that the short-range attraction that might bring them back together is far too weak to do this. As a result, the particles will continue to move rapidly apart with almost undiminished velocity, exploring beyond their previous positions. This process is shown in Fig. 2.2.

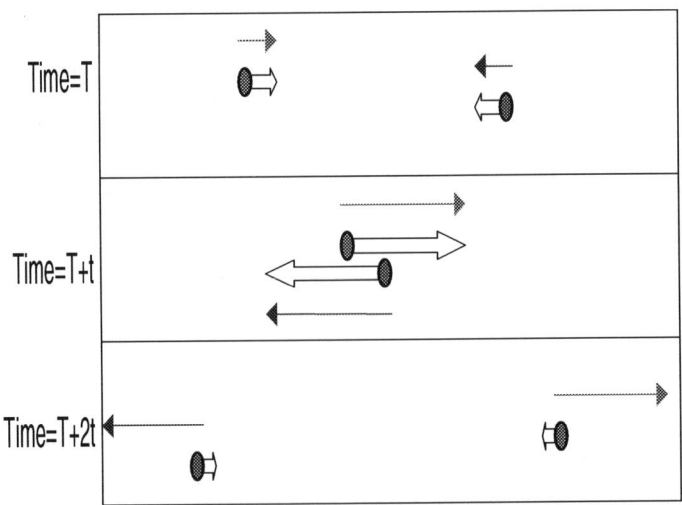

Fig. 2.2. The aliasing effect introduced by discrete evaluation

At time T, the separation between the particles is decreased to a value at which the magnitude of the short-range attraction (shown by broad arrows) is becoming significant. This augments the effect of their velocities (shown by

---

[1] Truly continuous evaluation is obviously impossible for any computer based PSO algorithm. Here the practical meaning of the word 'continuous' implies that the interval between evaluations is sufficiently short so that no particle can have moved a significant distance between evaluations

thin arrows) so that the particles move closer together at an increased rate. Remember that this force is considered to act unchanged for the next t time units. By time $T+t$ the particles are close and the short-range effect calculated now is large. As a result, the velocity of the particles increases substantially, almost entirely as a consequence of the short-range attraction. Again this force is considered to act unchanged for the next t time units. By time $T+2t$, when the next evaluation is made, the particles have passed each other and are so far apart that the short-range force calculated, which has changed direction, is weak. Consequently, the particles continue to diverge, retaining at $T+2t$ much of the velocity obtained as a result of the short-range forces calculated at time $T+t$. The short-range forces will continue to decrease as the particles move apart, leaving only the normal swarm factors to influence their future movement (in the absence of other close encounters).

If continuous evaluation could be used, the direction of the short-range force would reverse as soon as the particles passed each other and the still high values of short-range attraction would rapidly nullify the effects of the acceleration experienced as the particles converged. After the particles have passed they would slow as they separated, then come to rest before once again converging and repassing. Eventually the particles would enter into a stable limit cycle.

Using discrete evaluation the effect described might occur if two particles happen to pass closely in a region of indifferent fitness but is most likely to happen as particles converge on a single optimum. In this latter case the short-range force will have the desirable effect that some of the neighbourhood so engaged will be ejected with significant velocities, and thus reduce the number of those left to converge and therefore reduce the probability of further ejections.

Hence the effect of the short-range force on the normal settling behaviour of the PSO is self-limiting with some swarm particles being left to continue exploration in the local vicinity. Further discussion of such a short-range force can be found in [5].

2.2.3 Organising Ejected Particles into Waves

When the ejected particles still have knowledge of the fitness at their point of ejection, simply ejecting particles from the locations of a known optimum will only result in another optimum being found under very favourable circumstances. This is because this retained knowledge will tend to draw them back to that promotion point unless they happen upon an area with even better fitness before the effect of the global attraction cancels their short range force induced velocity. This problem is shown diagrammatically for one dimension in Fig. 2.3. Here a particle has been ejected from the right minimum but it is not assured that it will have sufficient momentum to reach the limited region in which the left minimum has a value less than the right minimum. As least, in one dimension the particle has a 50:50 chance of heading in the right direction. As the number of dimensions increases the probability that the particle will head exactly toward any small potentially useful good fitness region rapidly decreases toward zero. In practice, even for a small number of dimensions, the probability that an ejected particle will fall back to the last promotion point is virtually unity.

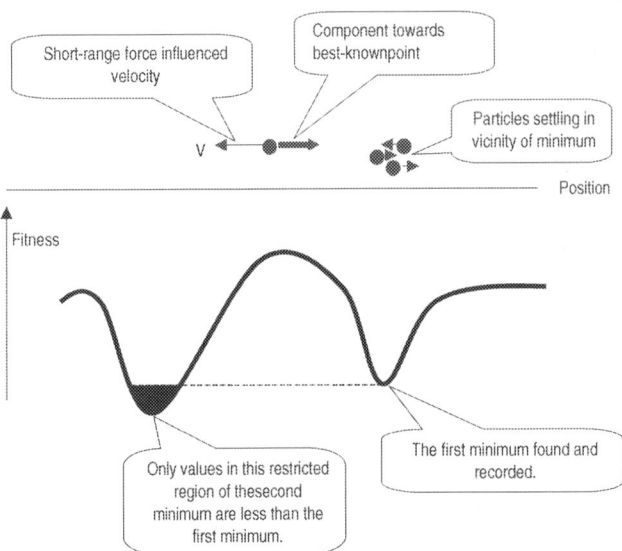

Fig. 2.3. The problem of finding a small region of better performance in one dimension when knowledge about the previous best performance point is retained

One solution is for ejected particles to 'forget' all about the optimum from which they were ejected. Assigning each particle to a wave, and treating each wave as a "sub swarm", can achieve this. Particles in a particular wave are only allowed to respond to the values of $\overline{B}$ reported by particles in that wave.

Initially all particles belong to wave zero. Every time a particle is ejected it is promoted by having its wave number increased to that of the highest numbered wave (creating a new wave if it is already in the highest numbered wave) and it starts to respond to the other particles in its new wave (if any). Since particles are commonly (but not always) ejected in pairs a wave will typically have an initial size of two. When two particles start a new wave (say wave N), the initial value of $\overline{B}$ will be the best position occupied by one of these particles. To reduce the probability that this will be a point associated with the optimum the particles are just leaving, every particle is required to move a user specified minimum distance (the *search scale*) away from its promotion position before it is allowed to inform other members of its wave on its current situation. This, together with the high ejection velocity, and an active repulsion from this particle's promotion points until it is at least a distance of *search scale* from all of its promotion points, sharply reduces the probability that particles will fall back to the optimum region associated with the wave they just left.

As most particles are promoted to successively higher waves, it is possible for some particles to get 'left behind' still circling an optimum found long ago but not getting close enough to other particles be ejected so as to join a new wave. As such these particles represent a wasted resource.

Cleaning up left over wave remnants can be achieved in two ways. Firstly, when a promotion event takes place that would leave only one particle in a wave, this last particle is also recruited to the wave this last promoted particle has just joined (but does not record this on its list of promotion points). Secondly, another inter-wave interaction is introduced. Once the value of $\overline{B}$ for wave N is better than the value of $\overline{B}$ for some earlier (lower numbered) wave, all remaining members of the earlier wave are promoted to wave N (but do not record this on their list of promotion points) and start to respond to wave $N's$ $\overline{B}$ value. Assuming that these just promoted particles do not uncover any region of interest while in transit (or move close enough to another particle so as to cause another promotion event), they will finally arrive in the region being explored by wave N.

This absorption of earlier (lesser performing) waves into later (higher performing) waves is more than just a remnant cleanup. The key is that this absorption only occurs into *better performing* later waves. This introduces an evolutionary pressure into the wave behaviour, with the result that quality of the maxima explored tends to increase with time, although not in a monotonic way.

2.2.4 When is a Particle Ejection a Promotion?

Ideally particles should be considered as ejected (and therefore promoted) when they come close while the wave is settling on an optimum. As a swarm settles the speed of the particles naturally deceases. The ratio of the velocity component introduced by the effect of the short-range force to the particle's other velocity components is a suitable measure with which an ejection can be detected. Promotion is deemed to have occurred any time this ratio exceeds a user specified value, called the *promotion factor*

2.2.5 Adding a Local Search

As each wave died (by losing its last particle or by having all its remaining members compulsorily promoted to a later wave that was outperforming it) a simple hill climbing[2] local search agent can be used to find the local optimum in the vicinity of the best position known to this wave. Directed random search was used as the local search agent for all results described later in this chapter.

Many optimisation heuristics combine an algorithm with coarse global search capabilities together with a suitable local search heuristic. While local search can play an important part in refining the optima found, it should be used sparingly lest the computational cost becomes too high. This extra expenditure of computing resource is ideally only warranted in the vicinity of an optimum, a condition that precludes its application to the basic particle swarm algorithm. The high probability that the best position known to a WoSP wave is, by its death, in the vicinity of an optimum makes its use only in the relatively few occasions when a wave dies both highly rewarding and computationally reasonable.

2.3 The WoSP Algorithm in Detail

At any time, each particle is in one of two different modes of behaviour as determined by its distance from the closest of all of the particle's previous promotion points. Only if this distance is more than the *scale search* parameter is the particle able to report its fitness and position to its wave. Only reporting particles are allowed to respond to their waves $\overline{B}$ value.

The algorithm proceeds as shown in pseudo code in algorithm 4.

Note that in equation 2.3 the attraction to the points $\overline{B}$ and $\overline{S}$ are now independent of the distance from these points. This encourages exploration by moderating the tendency to rush toward these positions. The constriction factor χ from equation 2.1 is not required here and so is omitted: however, the speed of any particle was clipped if necessary to a user specified maximum allowed speed.

Without the component $\hat{P}_t$ in step 7, experience shows that exploration tends to be concentrated on a hyper plane defined by the positions of the first few of the particle's promotion points. Adding the extra component $\hat{P}_t$ is one way to discourage this and encourages exploration throughout the problem space.

2.3.1 The Computation Cost of the WoSP Algorithm

The extra computational cost introduced to the basic swarm algorithm by the short range force alone and by the full WoSP algorithm can be calculated by timing a series of repeats with a fitness function that return a constant value. No swarm coalescing takes place under these conditions and the number of

---

[2] Hill climbing when maxima are being sort, hill descending when the object is to locate minima

Algorithm 4 The WoSP algorithm

1.   Particles are randomly positioned in problem space and given random velocities.
2.   The new position of each particle is first calculated, based on the position and velocity calculated in the previous iteration and assuming that the velocity remained unchanged for the time t between the iterations. The net short range force ($NSRF$) acting on each particle is calcuated.
3.   A check is made of the closest distance of each particle to any of its promotion points and its report status is updated as required.
4.   The fitness of each particle is calculated.
5.   Starting from the particle with the best fitness, and in descending order of fitness, those particles allowed to report do so to their wave, updating that waves best-known point as required.
6.   Each time a wave updates its best position a check is made to see if this in now fitter than the best position of some other lower numbered (earlier) wave. When this occurs, all members of earlier wave immediately join that latter wave without recording their current position as a promotion point.
7.   The velocity of every particle is now updated. Particles that are allowed to report update their velocity as:

$$\bar{V}_{T+t} = (M\bar{V}_T + rand \cdot G(\frac{\bar{X} - \bar{B}}{|\bar{X} - \bar{B}|t}) + rand \cdot L(\frac{\bar{X} - \bar{S}}{|\bar{X} - \bar{S}|t}) + NSRF) \qquad (2.3)$$

Particles that are not allowed to report, as they are too close to one or more of their promotion points, update their velocity as:

$$\bar{V}_{T+t} = (\bar{V}_T - G\hat{P}_c - L\hat{P}_t + NSRF) \qquad (2.4)$$

where $\hat{P}_C$ is the unit vector toward the closest previous promotion point and $\hat{P}_t$ is the unit vector in the direction of the smallest absolute component of $\hat{V}_T$.
8.   Every particle whose velocity component caused by the short-range force is more than *promote factor* times the vector sum of the other velocity components is promoted. It either joins the highest current number wave or, if it is already a member of the highest number wave, starts a new wave with a wave number of one higher. The position it was in when promoted is added to the particle's list of promotion points. If this promotion leaves a wave with only one member, this is also promoted as part of the process of cleaning up old waves. This compulsory recruited particle does not record its position as a new promotion point.
9.   If the best performance is adequate or enough time has passed without any change to *gbest* exit, else return to step 2.

promotions (when these are allowed to occur) is a function of the starting conditions. The average extra computation observed from 100 repeats, compared to the basic swarm algorithm, was about 55% for the short-range force only and just over 60% for the full WoSP algorithm. These values refer to the basic algorithms excluding the time for fitness evaluation. Since fitness assessment is often the dominant computational cost in real life problems, this means that for such problems the overhead introduced by the WoSP algorithm would be very small compared to the overall time for a conventional swarm algorithm.

2.3.2 Interactions between the WoSP Parameters

The promotion process require only one parameter, the *promotion factor* which is the minimum ratio of the SRF induced velocity component to the velocity excluding the SRF component needed to cause a promotion event. However, the short range force itself involves two parameters, the SRF_{power} and the constant K.

The values for these three parameters should be chosen so that a sufficient number of promotion events occur to permit exploration away from known optima, without so many occurring that the normal convergence of a swarm on optima is excessively impeded. Experimentation with numerous combinations of values shows that none of the values for these three parameters is critical. Choosing a high value for SRF_{power} so that the effect of the short-range force falls off very fast with distance alters the velocity spectrum of promoted particles so that, while the frequency of promotions decreases, the ejection velocity is typically higher encouraging aggressive exploration. However, experiments show that using a lower value of SRF_{power}, while taking longer, achieves substantially the same results.

Two parameters are involved when a particle is within *search scale* of its closest promotion point. These are *search scale* itself, which effectively sets the minimum inter optimum separation that the waves can be expected to readily identify as separate optima, and the extra repulsion weighting RW to apply to the particle to encourage it to aggressively explore.

All of these five parameters, except for *promote factor*, are functions of the dimensions of the problem space. A Cartesian distance of $x\sqrt{n}$ separates two particles a distance of x apart in each of n dimensions. As the dimensions of the problem increase, the average spacing between particles also increases. Thus, when changing from a problem in d_1 dimensions to the same problem in d_2 dimensions, the values of SRF_{factor} and of *search scale* should be altered by $\sqrt{\frac{d_2}{d_1}}$. As the number of dimensions increases, the increase in the average inter-particle separation suggests it may be advantageous to decrease the value of SRF_{power}. For the same reason it may be advantageous to increase the repulsion weighting RW

2.4 The Performance of the WoSP Algorithm

2.4.1 A Two Minimum Problem

Before considering the performance of the full WoSP algorithm, it is instructive to look at the results of adding the short-range force alone, that is without
the addition of waves.

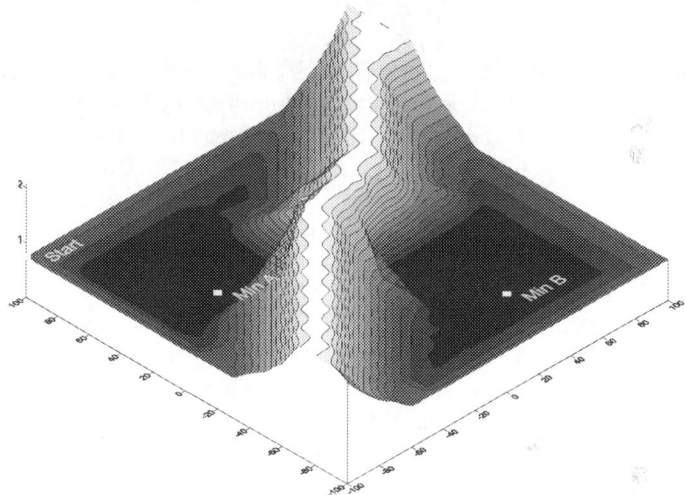

Fig. 2.4. A two dimensional fitness surface in which two minima are separated by
a poor fitness hill

As a first example consider the two-dimensional two minimum test fitness
surface shown in Fig. 2.4. This surface, which was explicitly defined point by
point rather than being generated from a function, contains two local minima,
A and B, separated by a high, poor fitness, ridge. Minimum B (value zero) is
marginally better than minimum A (value 0.1) while on the intervening ridge
the value climbs to a value of two. Points off the drawn map were assigned
a value equal to that of the closest on-map point. All swarm members were
initialised around the edge of the drawn map in the region in which -105
<X<-95 and 105 < Y < 95. Particles therefore were closer to minimum A
than to minimum B. The directions of the initial velocities were randomly
chosen with the magnitudes being assign a random value between specified
minimum and maximum values.

Swarm size	50
Global weighting G	0.5
Local weighting L	0.5
Momentum	0.9
Initial X position range	-95..-90
Initial Y position range	95..100
Minimum found	if an evaluation is made within a distance of 1 unit
Test terminated	if minimum B was not found after 2000 iterations

Table 2.1. The fixed test parameters

A number of tests were each repeated 1000 times, each test only differing in the values of some parameters. The maximum velocity that a particle could have was bounded to a value of *Max* times the maximum initial velocity, where *Max* is a user chosen parameter. This set the maximum distance that a particle could travel in an iteration, and therefore the maximum possible distance between fitness evaluations. The parameter values held constant across all tests are shown in Table 2.1 and the results from a chosen sample of these tests are shown in Table 2.2.

The surface and initial particle release region were specifically designed to make it difficult for a conventional swarm to find the better minimum, minimum B. As expected, without a short-range force, the swarm quickly found and settled on minimum A 100% of the time, taking an average of 21.5 iterations with a standard deviation of 8. This result was obtained with a maximum initial speed of 30.

The number of iterations taken when a short-range force was operational are shown in Table 2.2. It will be noticed that the shortest time occurs with the smallest maximum. As the maximum initial speed was increased, the probability that the particle would over fly the minimum in the period between evaluations also increased. As a result particles had to reverse and recross the minimum (possible several times) before it was observed, thus increasing the detection time. As the probability of a significant short-range force was low until the particles were congregating around the minimum, there was little chance for the speed to be accelerated beyond the initial velocity range. As a result the speed-limiting factor *Max* had no effect.

Finding the second minimum required that at least one (and probably many) significant short-range force events took place. Now the maximum distance that a particle can travel in one iteration (*max* times the maximum initial speed) sets the lowest resolution of the search. The distance between the two minima is just under 130 units. As the maximum distance per iteration rises significantly above this there is an increase in the number of iterations taken to find the minimum as more particles 'over fly' it.

Fig. 2.5 shows all the points evaluated in 400 iterations – the second minimum was found after 320 iterations. Note the curved paths at the top where

Table 2.2. The number of times the two minima were found and how long this took. Results derived from 1000 repeats for each set of parameter values for random initial particle positions

	Maximum:		First minimum:			Second minimum:		
Max	Initial speed	Distance / iteration	% times found	Average iterations	SD	% times found	Average iterations	SD
5	15	75	99.7	18.9	4.9	99.4	441.7	327.4
5	15	75	99.7	19	4.9	99.9	375.6	269.4
5	15	75	99.4	18.7	4.9	100	311.7	218.3
5	30	150	100	21.3	8.2	99	482.7	306.5
5	30	150	99.8	21.2	8.1	99.7	411.4	269
5	30	150	99.9	21.1	8.1	100	351.9	235.3
5	60	300	90.4	30.4	10.8	99	474.1	369.8
5	60	300	90.3	30.4	10.6	99.5	420	309.9
5	60	300	90.4	30.4	10.8	99.9	350.8	265.7
10	15	150	99.9	18.9	4.8	99.5	467.4	347
10	15	150	99.7	19	4.9	99.9	394.8	289.1
10	15	150	99.7	18.7	4.8	100	331.2	237.1
10	30	300	99.8	21.3	8	99.2	491.2	337.2
10	30	300	99.9	21.4	8.3	99.6	428	298.1
10	30	300	100	21.1	8	100	367.6	251
10	60	600	92.9	30.9	11.2	99.5	491.2	366.1
10	60	600	92.5	30.1	11.1	99.9	438.7	336.6
10	60	600	92.6	30.3	11.1	99.7	383	290.4

particles ejected into an 'unprofitable' region curved round and returned to the best known minimum, the one they had just left. Compare this with the paths left as particles divert from their path as they are attracted to the just discovered second minimum. The heavy concentration of points between the two minima results from all the particles in the vicinity of the first minimum being attracted to the new, better, minimum when it is found.

2.4.2 A Three Maximum Problem

The fitness surface shown in Fig. 2.6 has three maxima, labeled A, B and C whose fitness values and distances from the centre of the particle start region are shown in Table 2.3. This fitness surface is defined by the function in equation 2.5. The values of the constants that, when used with (2.5), produce the fitness surface shown in Fig. 2.6 are given in Table 2.4.

$$fit(x, y) = \sum_{i=0}^{2} \frac{H_i}{\left(1 + \sqrt{(x - x_i)^2 + (y - y_i)^2}\right)^{S_i}} \quad (2.5)$$

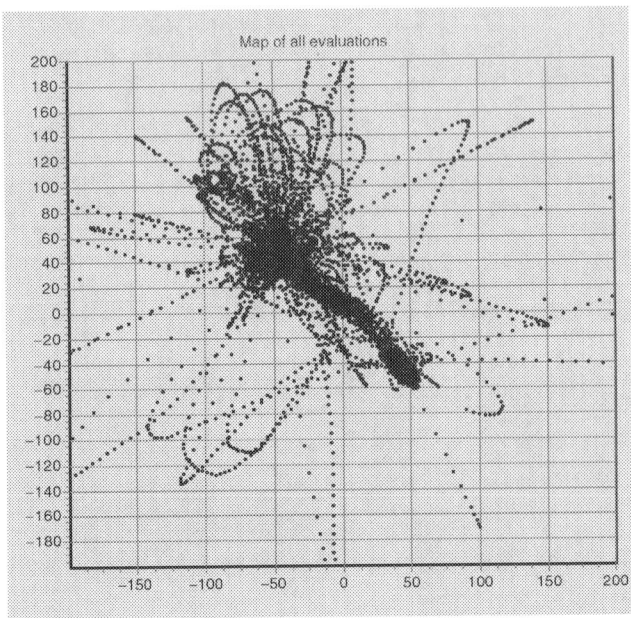

Fig. 2.5. All the points evaluated during a run of the WoSP algorithm on the surface shown in 2.4

Table 2.3. Maximum values and distances from the centre of the start circle for the problem surface shown in Fig. 2.6

Maximum	A	B	C
Value	8.176	8.732	8.518
Dist from average start position	272.8	428.7	513.4

Table 2.4. The constant values that when used in equation 2.5 produce the surface shown in Fig. 2.6

Peak	Index i	x_i	y_i	H_i	S_i
A	0	-40	-40	4.5	0.15
B	1	50	50	5	0.15
C	2	120	75	4	0.2

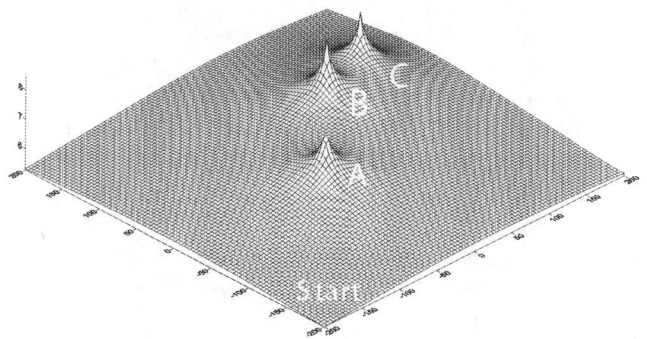

Fig. 2.6. The two-dimensional three maxima fitness surface

The swarm particles were initialised from random positions within a restricted region as indicated by the region labeled 'Start' (at the front of the figure). This point was chosen to be far from any maximum. The closest local maximum to this start region (maximum A) is the poorest of the three. The second closest maximum (maximum B) is the highest. This test was designed not only to see if particles would move from maximum A to maximum B (a better maximum) but also to examine how well they would explore beyond maximum A and investigate maximum C which was of intermediate fitness[3].

This problem was designed so that the performance of the algorithm with just the short-range force was essentially the same as for the basic PSO. This is because the probability of a particle ejected from maximum A encountering the limited region round maximum B that has a higher fitness than maximum A before the global component returns it to maximum A is very small.

The key parameters used are listed in Table 2.5 and the results from 1000 independent trials of both basic PSO and for WoSP are presented in Table 2.6. Note that, as far as finding the first maximum is concerned, the basic swarm and WoSP have the same results. However, the basic

---

[3] Although, knowing the maxima in this problem, it is clear that exploring from maximum B to maximum C is of no practical purpose as C is a lesser maximum, this information would not be available *a priori* for a real problem. Then the ability to leave a fit point and go on to explore a point of lower fitness could be an important step on the way to finding a further point with a fitness better than any yet explored

swarm does not explore further, with the result that just over 87% of the time it settles for the poorest of the three maxima. The one case in which the basic swarm found two maxima appears to be the result of the swarm happening to split into two and starting to coalesce on both maxima at almost the same time. The particles converging on maximum A met the criterion for 'found' just before the particles converging on maximum B. The WoSP algorithm, on the other hand, while again overwhelmingly identifying maximum A first continues to explore and finds the global maximum B every time, often exploring further and finding maximum C as well.

Table 2.5. The key SRF and wave parameter values used

SRF coefficient	0.01	Search scale	50
SRF power	2	Promotion factor	10

While it finds at least two maxima on every occasion, it only found all three maxima 78.8% of the time before the run was terminated after 1000 iterations. The basic PSO took just over 50 iterations (on average) to find the first (and, apart from one case, only) maximum. The WoSP algorithm took almost the same time to find the first maximum, but took some 800 iterations on average to find all three maxima (when it did in fact find all three).

Table 2.6. The relative performance of the basic PSO and WoSP algorithms

	Basic Swarm			WoSP		
Maximum	A	B	C	A	B	C
Found first	783	217	0	783	217	0
Found second	0	1	0	3	781	216
Found third	0	0	0	21	2	765
Total found	783	218	0	807	1000	981

2.4.3 A Dual Cluster Problem

The dual cluster problem space was designed to investigate the effect of the *search scale* parameter on the performance of the WoSP algorithm. The space consists of two clusters of minima, each cluster consisting of six lesser minima surrounding a central better minimum as shown in Fig. 2.7.

In this space the fitness of a particle at some position is the minimum of the score values from that position to each of the 14 defined points. The score

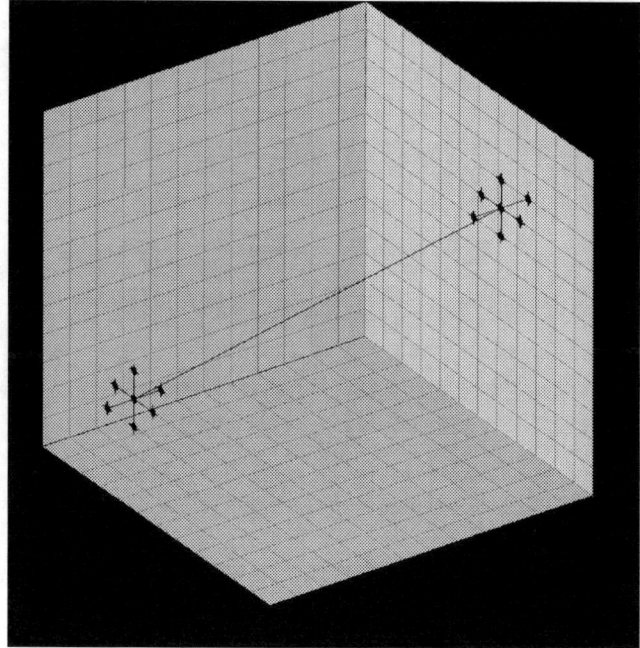

Fig. 2.7. The 14 minima, two cluster problem space

value of point X is the cartesian distance from the particle to X plus the floor value of point X: the position and floor values of the minima are given in Table 2.7.

This produces two equal global minima with values of zero at points A- and A+, together with twelve other local minima (one at each of the other defined points) each with a value of five. Each of these local minima is a distance of exactly 50 units from one of the global minima, pairs of local minima within a cluster are separated by either just under 71 or by exactly 100 units. The two global maxima were approximately 700 units apart.

The results obtained from 100 independent runs at each of five search scales are shown in Table 2.8. The figures are number of times this maximum (or group of maxima) was found in 100 independent trials. The first search scale was chosen to be less than the closest spacing while the rest were chosen to be at approximately one of the spacings between minima.

Examination of these result for the dual cluster problem space reveals the following:

- No wave ever converges at a point that is not in the near vicinity of a minimum.
- Although a particle was prohibited from reporting when within search scale of any of it's promotion points, minima can be re-explored by different

Table 2.7. The position and floor values of each of the 14 minima

	x	y	z	Floor
A-	-200	-200	-200	0
B-	-150	-200	-200	5
C-	-250	-200	-200	5
D-	-200	-150	-200	5
E-	-200	-250	-200	5
F-	-200	-200	-150	5
G-	-200	-200	-250	5
A+	200	200	200	0
B+	150	200	200	5
C+	250	200	200	5
D+	200	150	200	5
E+	200	250	200	5
F+	200	200	150	5
G+	200	200	250	5

Table 2.8. The times each minimum, or selected combinations of minima, were found in 100 WoSP runs

	Search scale				
Minimum	30	50	70	100	700
A-	77	61	56	97	65
B-	145	31	16	22	5
C-	190	15	29	33	17
D-	147	25	23	19	14
E-	214	19	27	36	30
F-	147	38	18	19	8
G-	198	18	35	34	23
A+	66	50	61	95	49
B+	127	11	13	19	16
C+	163	38	33	21	35
D+	143	12	21	13	16
E+	194	36	23	27	34
F+	126	12	22	21	15
G+	182	33	17	42	34
Any other point	0	0	0	0	0
A- or A+ found	100	100	98	100	99
A- and A+ found	28	8	12	60	15
A- found > once	8	0	1	12	0
A+ found > once	5	2	3	13	0

waves. This can occur since the particles in a particular wave may have reached that wave with different wave membership histories. This occurs as a promoted particle joins the highest number wave existing at the time of promotion, creating this wave if necessary. Thus particles in a wave can have quite different promotion histories. Once a particle that has not previously explored in the region of some optimum A reaches this region it can report to the wave, other particles that have already explored this region are none the less now attracted to it and, although they cannot report on it, will again converge on it and be promoted from it. This is an important feature of the algorithm: without it an optimum could only be used as a base for further exploration once.

- With a search scale less than the smallest inter-minimum spacing, there is substantial re-exploration, especially of the 'lesser' minima. As the search scale increases, this decreases but does not totally cease for the reason given above.
- The algorithm is effective at finding at least one or the two equal global minima for all values of search scale, but the probability of finding both increases as the search scale is itself increased as this prevents so much effort being spent on examining the 'lesser' minima. However, once the search scale approaches the separation between the two global minima, finding one will tend to preclude finding the second unless this is the very next minimum found after the first global one.
- Note how the two global minima $A-$ and $A+$ may be found more than once: if the search scale is small there is much exploration of all minima and so the particles of a particular wave may have many different wave membership histories. This, of course, can allow re-exploration of 'lesser' minima too, but as the search scale is increased the number of lesser minima available to be found is decreased.

2.4.4 A Problem with 8^{30} Maxima

The results for a simple problem presented in Table 2.8 do not illustrate the full potential of the WoSP algorithm. This becomes more obvious from consideration of the results obtained seeking the maximum for Schwefel's function in 30 dimensions:

$$f = \sum_{i=1}^{30} x_i \sin(\sqrt{|x_i|}) \tag{2.6}$$

$$where \quad |x_i| \leq 500.$$

This function is suitable for exploring behaviour of algorithms in multi-maxima situations as not only does it have multiple maxima, but also the highest and second highest maxima are separated by a number of lesser maxima. Each individual dimension's contribution to the overall value as shown in Fig. 2.8. Table 2.9 shows the positions and values of each maximum, rounded to the nearest integer.

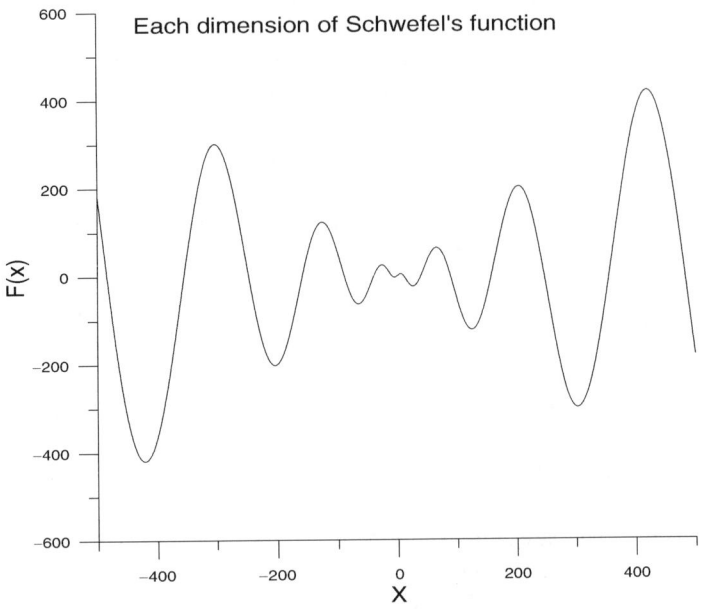

Fig. 2.8. Schwefel's function in 1 dimension

Table 2.9. The eight maxima per dimension of Schwefel's function, values rounded to the nearest integer

Maximum	X	f(X)
A	421	419
B	-303	301
C	204	202
D	-500	181
E	-125	123
F	66	64
G	-26	24
H	5	4

Each dimension is identical as shown in Fig. 2.8, thus allowing the absolute maximum to be readily calculated for any number of dimensions. This, together with the significant separation between the maximum and second highest peaks, makes this function suitable for exploratory work to find the maximum in any number of dimensions.

In 30 dimensions, Schwefel's function has 8^{30} maxima, with one global maximum of 12569.5 when $x_i=420.9687$ for all 30 values of i. The constraint that $|x_i| \leq 500$ for all values of i was hard coded; any particle reaching this limit underwent a fully elastic rebound.

The performance of a basic swarm algorithm on this problem from 100 trials (using the same basic values as shown in Table 2.10 except for the maximum number of iterations which was set to 100,000) is shown in Fig. 2.9. Note that the best value found in any trial was still less than 10,000.

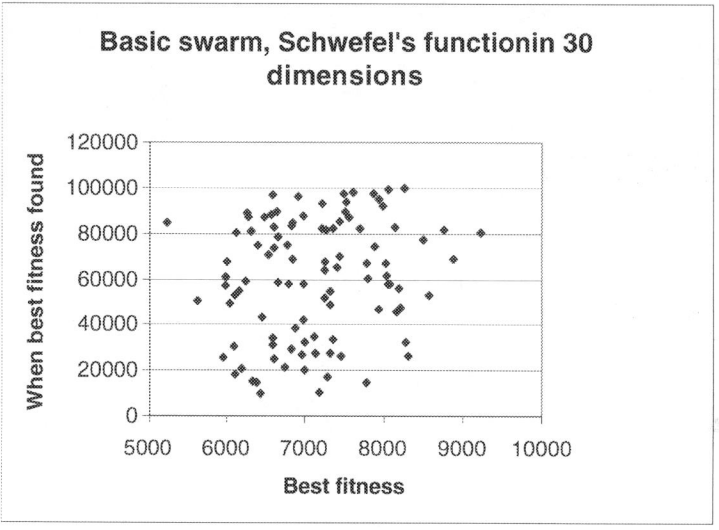

Fig. 2.9. The performance of 100 independent trials of the basic swarm algorithm on Schwefel's function in 30 dimensions

Using the parameter values shown in Table 2.10, in a series of 100 trials each for 200,000 iterations (an iteration consists of all particles making one velocity update), the global maximum was found 41 times by the WoSP algorithm. On average this best position was one of more than 100 explored during the run and was found after about 118,000 iterations. A second set of runs, identical except for the duration, which was set to 2,000,000 iterations, showed a slight performance change, but one that was insignificant when compared with the order of magnitude increase in computing cost.

As each wave died (lost its last particle or had all its members compulsorily promoted to a later wave that was outperforming it) a simple directed random search hill climbing local search agent was used to find the local optimum in the vicinity of the best position known to this wave as described before.

Table 2.10. The parameter values used for Schwefel's function in 30 dimensions

Parameter	Value
Number of particles	30
Maximum number of iterations	200,000
Total number of evaluations	6,000,000
Momentum	0.95
B global best factor	0.9
Normal$L$ local best factor	0.5
L if within search scale of a promotion point	20
Search scale	500
Promote factor	2

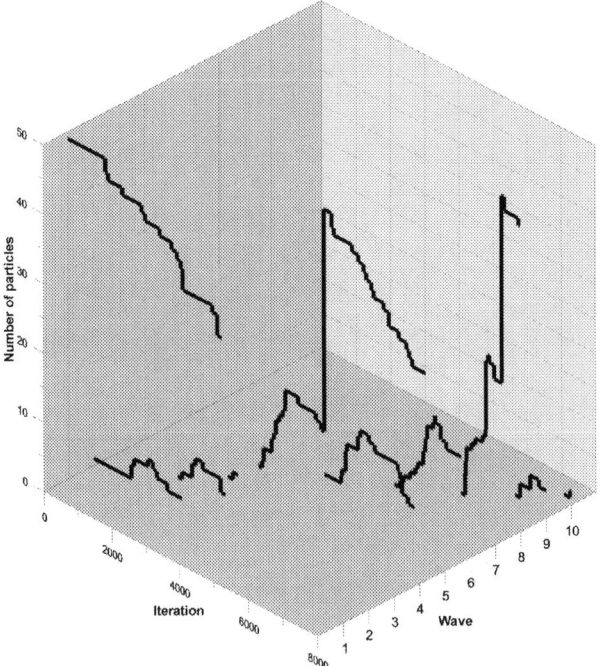

Fig. 2.10. A history of the number of particles in each wave during the first 8000 WoSP iterations (Schwefel's function in 30 dimensions)

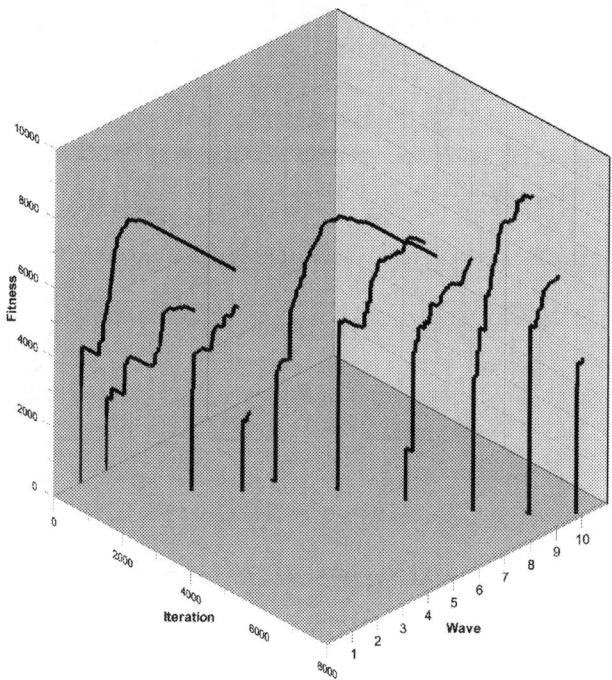

Fig. 2.11. A history the best fitness discovered by each wave during the first 8000 WoSP iterations (Schwefel's function in 30 dimensions)

This extra expenditure of computing resource is ideally only warranted in the vicinity of an optimum, a condition that precludes its application to the basic particle swarm algorithm. The high probability that each wave has by its death investigated in the vicinity of an optimum makes its use in these few positions both highly rewarding and computationally reasonable. Fig. 2.10 shows the number of particles in each wave, and Fig. 2.11 the best fitness yet found by each wave, for each of the first 8000 iterations of a run in which the WoSP algorithm was finding the maxima of Schwefel's function in 30 dimensions. Note that waves 8 to 10 have not yet died by the time the plots finish. Note how later waves (e.g. wave 8) absorb the particles from earlier waves whose fitness they exceed. Also note that waves exploring poorer regions tend persist for shorter times than waves exploring better regions.

Fig. 2.12 shows the best fitness achieved by each wave as it died (or by 200,000 iterations when the run was terminated) for a typical run. In each case, at death or termination, a local search agent used directed random search to explore the local optimum. It is the fitness found by this local search agent that is assigned to the wave and shown. As waves are numbered sequentially as they are created, the wave number is an approximate guide to the position in the run when this wave was created. In this particular run 98 different

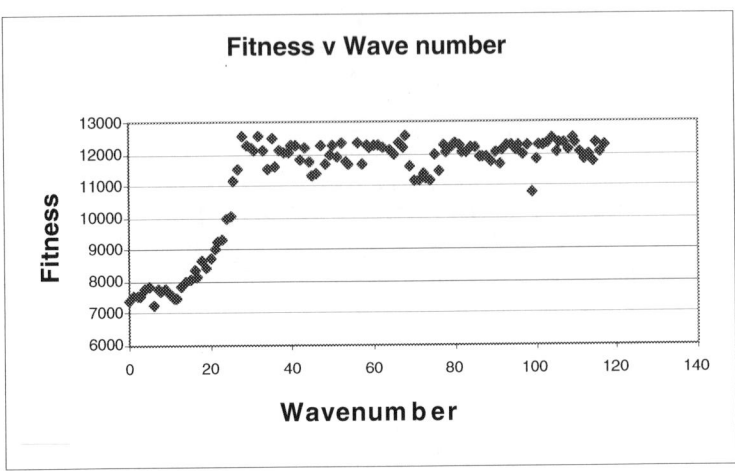

Fig. 2.12. The best fitness achieved by each wave during a typical run (Schwefel's function in 30 dimensions)

waves reported on the maxima they explored. Three maxima, one of which was the global maximum, were explored three times while nine others were explored twice. All other 86 maxima were only explored once.

The absorption by a later wave of the particles of any earlier wave that it is outperforming is the reason for the fairly steady improvement leading up to the first exploration of the global maximum by wave twenty-eight. This absorption process is effectively a survival of the fittest with the absorbed particles moving to join the particles already in the absorbing wave. Once there they can explore the immediate vicinity but they, like all particles there, may also be promoted again thus using this as a springboard for further exploration seeking even better maxima. Thus an evolution process is taking place with these waves, which provides the early sharp improvement in the fitness of maxima explored. Once the global maximum has been found evolution can only operate beneficially on waves that are exploring sub-global maxima.

Particles are actively rejected if they again approach a position from which they have been promoted and are prohibited from providing feedback to their wave if within *search scale* of such a position. The particles of wave 28 are, in time, promoted into later waves but cannot notify their wave of any fitness in the vicinity of the global maximum. The limited re-exploration that occurs must be initiated by some particle that has not been promoted from a position close to the global maximum and is thus able to report from the vicinity of the global maximum: the other particles in its wave can still respond to its reports even though they themselves cannot report.

Re-exploration could be totally prohibited if each particle, instead of maintaining a list of its own promotion points, had access to a global list of all promotion points. However, this would likely be of disadvantage as it would make the progress of waves in the problem space a series of linear steps from maximum to maximum without (as now) allowing for forking to occur and one maximum to be used as a base for exploring two or more other maxima.

2.4.5 A Problem with 8^{100} Maxima

A further series of experiments was undertaken involving Schwefel's function in 100 dimensions. This problem has a total of 8^{100} maxima with a single global maximum when each of the 100 dimensions is set to approximately 421 (A in Table 2.9).

The parameters SRF_{factor} and *search scale* were adjusted as described in Section 2.3.2 from the values that had proved best in 30 dimensions to 913 and 9130 respectively. Tests using values of SRF_{power} of 2.5, 3 and 3.5 were performed.

The global maximum was not found during any of the 30 repeats with each of these values, but for $SRF_{power}=3$ the best result each run had on average 97 dimensions correct (A in Table 2.9), with value of the remaining three being either B or C. The non-optimal dimension values were not the same, nor in the same position, each time.

Unlike the function in 30 dimensions, for which the best value of SRF_{power} was 3.5, in 100 dimensions the best value was 3. The average best fitness was 40390 for $SRF_{power} =2.5$, 41046 for $SRF_{power} =3$ and 35768 for $SRF_{power} =3.5$. The best possible fitness is 41898. These results were obtained from runs limited by practical considerations to 500,000 iterations. Altering the SRF_{power} also changed the average time at which the best result was found. Lower values of SRF_{power} resulted in the best result being found earlier.

2.5 Comparison to Other Approaches

A number of variations to the basic PSO algorithm have been proposed that are intended to either promote convergence to a number of sub-swarms so that a range of optima may be explored in parallel and / or permit the tracking of optima of a problem domain with a temporal component. [7] and [8] discuss two such methods and contain a review of a number of other approaches.

These PSO versions use techniques already proven in niching genetic algorithms, such as fitness sharing, speciation and fitness function modification. These techniques aim to explore several optima simultaneously, with the number exploring a particular optimum being approximately proportional to the relative fitness of the optimum.

While it is true that the WoSP algorithm may result in several optima being explored in parallel, the main emphasis in WoSP is on the sequential exploration of optima. Given that the size of a practical swarm must always be limited, and that the absolutely maximum number of optima that can be concurrently explored can never exceed the swarm size, it follows that a niching PSO must always be limited in the number of optima that it can explore. The WoSP algorithm has, in principle, no such constraint.

2.6 Constraint Handling

All the trial fitness functions described in this chapter share one common feature - all points on the surface have a valid fitness. In real life this is not always so, consider optimising a complex process, some combinations of parameters may result in invalid solutions. The classical PSO algorithm has a capacity to handle such situations. As long as neither *gbest* nor *lbest* are allowed to be updated with a position that corresponds to an invalid solution, search will automatically tend to concentrate on regions in which solutions are valid. The provision of momentum will encourage particles to traverse regions that correspond to invalid fitnesses. The WoSP algorithm shares all these characteristics with the PSO algorithm. In addition, since the short range force is valid everywhere, the rapid ejection mechanism has the ability to result in a rapid traversal of regions. All that is required is that the algorithm is amended so that the conditions that must be met in order for a particle to be allowed to report to their wave be expanded to include the condition that the fitness obtained corresponds to a valid solution. Positions that correspond to invalid solutions can be added to the rejection list of a particle, although this may result in the list becoming excessively long. Preliminary work has shown that such an approach works well for fitness surfaces with a minority of regions of non-validity, be these regions bounded or unbounded. The performance on fitness surfaces for which the majority of positions correspond to invalid solutions is less successful. Further work currently underway aims to investigate and hopefully improve the constraint handling capabilities of WoSP.

2.7 Concluding Remarks

This chapter introduces a version of the particle swarm optimisation algorithm intended to be efficient when exploring problem spaces with multiple optima. As no additional fitness evaluations are required for the WoSP algorithm compared to basic PSO, the additional computational cost of the behaviour modification is likely to be relatively small when compared to the overall computation involved, especially for problems with complex fitness functions.

The results obtained on a simple contrived three-maximum problem clearly show that the WoSP algorithm is able to escape from local sub-optima and continue to search for other optima.

The dual cluster problem space results are instructive in showing both the effect of the choice of search scale on this simple problem but also how re-exploration of previously explored optima is important in preventing the algorithm from being too greedy by allowing more than one wave of exploration to be initiated from one place.

Results obtained from the more challenging Schwefel's function are most instructive. When one considers that in the 30 dimension case, 6×10^6 evaluations were done during the 200,000 iterations for a 41% chance of finding the best of approximately 1.2×10^{27} maxima, the performance of this technique on this problem is quite remarkable and a testament to the power of combining swarm exploration of each maximum with an evolutionary driven search for further maxima. In 100 dimensions, the absolute best of the $2*10^{90}$ maxima were never able to be found within the $15*10^6$ evaluations performed, but the best fitness found each run was reliably in the top $2*10^{-82}$%. The reports on the locations and fitness's of other maxima obtained for these problems are a bonus that may be of considerable use in practical problems, such as scheduling.

The regular spacing of the maxima in Schwefel's function may have particularly suited the WoSP algorithm, but the results are sufficiently encouraging to augur well for other problem domains.

References

1. R. Brits, A.P. Engelbrecht, F. van den Bergh, A Niching Particle Swarm Optimizer, Proceedings of the 4th Asia-Pacific Conference on Simulated Evolution and Learning 2002 (SEAL 2002), Singapore. pp. 692-696, 2002
2. R.C. Eberhart,P. Dobbins and P. Simpson, Computational Intelligence PC Tools, Academic Press, Boston, USA 1996
3. R. Eberhart and Y. Shi, Comparing Inertia Weights and Constriction Factors in Particle Swarm Optimisation, Proceedings of the 2000 Congress on Evolutionary Computation, pp. 84-88, 2000
4. T. Hendtlass, A Combined Swarm Differential Evolution Algorithm for Optimization Problems. Lecture Notes in Artificial Intelligence, Vol 2070, pp. 374-382, Springer, Berlin. 2001
5. T. Hendtlass and T. Rodgers, Discrete Evaluation and the Particle Swarm Algorithm, Proceedings of Complex04, Cairns, Australia, pp. 14-22, 2004
6. J. Kennedy and R.C. Eberhart, Particle Swarm Optimization, Proc. IEEE International Conference on Neural Networks, Perth Australia, IEEE Service Centre, Piscataway NJ USA IV:pp. 1942-1948. 1995
7. J. Kennedy and R.C. Eberhart, The Particle Swarm: Social Adaptation in Information-Processing Systems, Chapter 25 in New Ideas in Optimization. Corne D., Dorigo M., and Glover F. (Editors) McGraw-Hill Publishing Company, England, ISBN 007 709506 5, 1999
8. D. Parrot and X. Li, A Particle Swarm Model for Tracking Multiple Peaks in a Dynamic Environment using Speciation, Proceeding of the 2004 Congress on Evolutionary Computation (CEC'04), p.98 - 103, 2004

9. S. Rana, L.D. Whitley and R. Cogswell, Searching in the presence of noise, in Parallel Problem Solving from Nature (PPSN IV), Lecture Notes on Computer Science 1141, Springer-Verlag, Berlin pp 198-207, 1996

3

Grammatical Swarm: A Variable-Length Particle Swarm Algorithm

Michael O'Neill[1], Finbar Leahy[2], and Anthony Brabazon[1]

[1] University College Dublin, Belfield, Dublin 4, Ireland.
   m.oneill@ucd.ie, anthony.brabazon@ucd.ie
[2] University of Limerick, Limerick, Ireland.
   finbarleahy@gmail.com

This chapter examines a variable-length Particle Swarm Algorithm for Social Programming. The Grammatical Swarm algorithm is a form of Social Programming as it uses Particle Swarm Optimisation, a social swarm algorithm, for the automatic generation of programs. This study extends earlier work on a fixed-length incarnation of Grammatical Swarm, where each individual particle represents choices of program construction rules, where these rules are specified using a Backus-Naur Form grammar. A selection of benchmark problems from the field of Genetic Programming are tackled and performance is compared to that of fixed-length Grammatical Swarm and of Grammatical Evolution. The results demonstrate that it is possible to successfully generate programs using a variable-length Particle Swarm Algorithm, however, based on the problems analysed it is recommended that the simpler bounded Grammatical Swarm be adopted.

3.1 Introduction

One model of social learning that has attracted interest in recent years is drawn from a swarm metaphor. Two popular variants of swarm models exist, those inspired by studies of social insects such as ant colonies, and those inspired by studies of the flocking behavior of birds and fish. This study focuses on the latter. The essence of these systems is that they exhibit flexibility, robustness and self-organization [2]. Although the systems can exhibit remarkable coordination of activities between individuals, this coordination does not stem from a 'center of control' or a 'directed' intelligence, rather it is self-organizing and emergent. Social 'swarm' researchers have emphasized the role of social learning processes in these models [6, 7]. In essence, social behavior

M. O'Neill et al.: *Grammatical Swarm: A Variable-Length Particle Swarm Algorithm*, Studies in Computational Intelligence (SCI) **26**, 59–74 (2006)
www.springerlink.com

helps individuals to adapt to their environment, as it ensures that they obtain access to more information than that captured by their own senses.

This paper details an investigation examining a variable-length Particle Swarm Algorithm for the automated construction of a program using a Social Programming model. The performance of this variable-length Particle Swarm approach is compared to its fixed-length counterpart [15, 17] and to Grammatical Evolution on a number of benchmark problems. In the Grammatical Swarm (GS) methodology developed in this paper, each particle or real-valued vector, represents choices of program construction rules specified as production rules of a Backus-Naur Form grammar.

This approach is grounded in the linear Genetic Programming representation adopted in Grammatical Evolution (GE) [18], which uses grammars to guide the construction of syntactically correct programs, specified by variable-length genotypic binary or integer strings. The search heuristic adopted with GE is a variable-length Genetic Algorithm. A variable-length representation is adopted as the size of the program is not known a-priori and must itself be determined automatically. In the GS technique presented here, a particle's real-valued vector is used in the same manner as the genotypic binary string in GE. This results in a new form of automatic programming based on social learning, which we dub *Social Programming*, or *Swarm Programming*. It is interesting to note that this approach is completely devoid of any crossover operator characteristic of Genetic Programming.

The remainder of the paper is structured as follows. Before describing the Grammatical Swarm algorithm in Section 3.4, introductions to the salient features of Particle Swarm Optimization (PSO) and Grammatical Evolution (GE) are provided in Section 3.2 and Section 3.3 respectively. Section 3.5 details the experimental approach adopted and results, and finally Section 3.6 details conclusions and future work.

3.2 Particle Swarm Optimization

In the context of PSO, a swarm can be defined as 'a population of interacting elements that is able to optimize some global objective through collaborative search of a space.' [6](p. xxvii). The nature of the interacting elements (particles) depends on the problem domain, in this study they represent program construction rules. These particles move (fly) in an n-dimensional search space, in an attempt to uncover ever-better solutions to the problem of interest. Each of the particles has two associated properties, a current position and a velocity. Each particle has a memory of the best location in the search space that it has found so far (p_{best}), and knows the best location found to date by all the particles in the population (or in an alternative version of the algorithm, a neighborhood around each particle) (g_{best}). At each step of the algorithm, particles are displaced from their current position by applying a velocity vector to them. The velocity size / direction is influenced by the velocity in the previous iteration of the algorithm (simulates 'momentum'), and the location of a particle relative to its p_{best} and g_{best}. Therefore, at each step, the size and direction of each particle's move is a function of its own history (experience), and the social influence of its peer group.

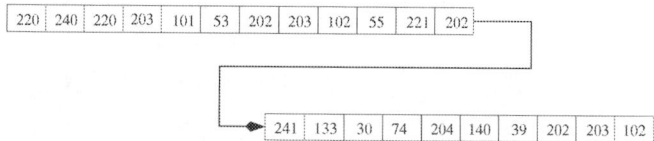

Fig. 3.1. An example GE individuals' genome represented as integers for ease of reading.

A number of variants of the particle swarm algorithm (PSA) exist. The following paragraphs provide a description of a basic continuous version of the algorithm.

i. Initialize each particle in the population by randomly selecting values for its location and velocity vectors.
ii. Calculate the fitness value of each particle. If the current fitness value for a particle is greater than the best fitness value found for the particle so far, then revise p_{best}.
iii. Determine the location of the particle with the highest fitness and revise g_{best} if necessary.
iv. For each particle, calculate its velocity according to equation 3.1.
v. Update the location of each particle according to equation 3.3.
vi. Repeat steps ii - v until stopping criteria are met.

The update algorithm for particle i's velocity vector v_i is:

$$v_i(t+1) = (w * v_i(t)) + (c_1 * R_1 * (p_{best} - x_i)) + (c_2 * R_2 * (g_{best} - x_i)) \quad (3.1)$$

where

$$w = wmax - ((wmax - wmin)/itermax) * iter \quad (3.2)$$

In equation 3.1, p_{best} is the location of the best solution found to-date by particle i, g_{best} is the location of the global-best solution found by all particles to date, c_1 and c_2 are the weights associated with the p_{best} and the g_{best} terms in the velocity update equation, x_i is particle i's current location, and R_1 and R_2 are randomly drawn from U(0,1). The term w represents a momentum coefficient which is reduced according to equation 3.2 as the algorithm iterates. In equation 3.2, $itermax$ and $iter$ are the total number of iterations the algorithm will run for, and the current iteration value respectively, and $wmax$ and $wmin$ set the upper and lower boundaries on the value of the momentum coefficient. The velocity update on any dimension is constrained to a maximum value of $vmax$. Once the velocity update for particle i is determined, its position is updated (equation 3.3), and p_{best} is updated if necessary (equations 3.4 & 3.5).

$$x_i(t+1) = x_i(t) + v_i(t+1) \tag{3.3}$$

$$y_i(t+1) = y_i(t) \text{ if, } f(x_i(t)) \leq f(y_i(t)) \tag{3.4}$$

$$y_i(t+1) = x_i(t) \text{ if, } f(x_i(t)) > f(y_i(t)) \tag{3.5}$$

After the location of all particles have been updated, a check is made to determine whether g_{best} needs to be updated (equation 3.6).

$$\hat{y} \in (y_0, y_1, ..., y_n) | f(\hat{y}) = \max (f(y_0), f(y_1), ..., f(y_n)) \tag{3.6}$$

3.3 Grammatical Evolution

Grammatical Evolution (GE) is an evolutionary algorithm that can evolve computer programs in any language [18, 19, 20, 21, 22], and can be considered a form of grammar-based genetic programming. GE has enjoyed particular success in the domain of Financial Modelling [3] amongst numerous other applications including Bioinformatics, Systems Biology, Combinatorial Optimisation and Design [16, 13, 5, 4]. Rather than representing the programs as parse trees, as in GP [8, 9, 1, 10, 11], a linear genome representation is used. A genotype-phenotype mapping is employed such that each individual's variable length binary string, contains in its codons (groups of 8 bits) the information to select production rules from a Backus Naur Form (BNF) grammar. The grammar allows the generation of programs in an arbitrary language that are guaranteed to be syntactically correct, and as such it is used as a generative grammar, as opposed to the classical use of grammars in compilers to check syntactic correctness of sentences. The user can tailor the grammar to produce solutions that are purely syntactically constrained, and can incorporate domain knowledge by biasing the grammar to produce very specific forms of sentences. BNF is a notation that represents a language in the form of production rules. It is comprised of a set of non-terminals that can be mapped to elements of the set of terminals (the primitive symbols that can be used to construct the output program or sentence(s)), according to the production rules. A simple example BNF grammar is given below, where <expr> is the start symbol from which all programs are generated. These productions state that <expr> can be replaced with either one of <expr><op><expr> or <var>. An <op> can become either +, -, or *, and a <var> can become either x, or y.

```
<expr> ::= <expr><op><expr>  (0)
         | <var>             (1)
  <op> ::= +                 (0)
         | -                 (1)
         | *                 (2)
 <var> ::= x                 (0)
         | y                 (1)
```

The grammar is used in a developmental process to construct a program by applying production rules, selected by the genome, beginning from the start symbol of the grammar. In order to select a production rule in GE, the next codon value on the genome is read, interpreted, and placed in the following formula:

$$Rule = c \mathbin{\%} r$$

where $\%$ represents the modulus operator, c is the codon integer value, and r is the number of rules for the current non-terminal of interest.

Given the example individual's genome (where each 8-bit codon is represented as an integer for ease of reading) in Fig.3.1, the first codon integer value is 220, and given that we have 2 rules to select from for <expr> as in the above example, we get 220 % 2 = 0. <expr> will therefore be replaced with <expr><op><expr>.

Beginning from the the left hand side of the genome, codon integer values are generated and used to select appropriate rules for the left-most non-terminal in the developing program from the BNF grammar, until one of the following situations arise: (a) A complete program is generated. This occurs when all the non-terminals in the expression being mapped are transformed into elements from the terminal set of the BNF grammar. (b) The end of the genome is reached, in which case the *wrapping* operator is invoked. This results in the return of the genome reading frame to the left hand side of the genome once again. The reading of codons will then continue unless an upper threshold representing the maximum number of wrapping events has occurred during this individuals mapping process. (c) In the event that a threshold on the number of wrapping events has occurred and the individual is still incompletely mapped, the mapping process is halted, and the individual assigned the lowest possible fitness value. Returning to the example individual, the left-most <expr> in <expr><op><expr> is mapped by reading the next codon integer value 240. This codon is then used as follows: 240 % 2 = 0 to become another <expr><op><expr>. The developing program now looks like <expr><op><expr><op><expr>. Continuing to read subsequent codons and always mapping the left-most non-terminal the individual finally generates the expression y*x-x-x+x, leaving a number of unused codons at the end of the individual, which are deemed to be introns and simply ignored. A full description of GE can be found in [18], and some more recent developments are covered in [3, 14].

3.4 Grammatical Swarm

Grammatical Swarm (GS) adopts a Particle Swarm learning algorithm cou-
pled to a Grammatical Evolution (GE) genotype-phenotype mapping to generate programs in an arbitrary language [15]. The update equations for the
swarm algorithm are as described earlier, with additional constraints placed
on the velocity and particle location dimension values, such that maximum
velocities $vmax$ are bound to ± 255, and each dimension is bound to the range
[0,255] (denoted as $cmin$ and $cmax$ respectively). Note that this is a continuous swarm algorithm with real-valued particle vectors. The standard GE
mapping function is adopted, with the real-values in the particle vectors being rounded up or down to the nearest integer value for the mapping process.
In contrast to earlier studies on GS this study adopts variable-length vectors.
A vector's elements (values) may be used more than once if wrapping occurs,
and it is also possible that not all dimensions will be used during the mapping process if a complete program comprised only of terminal symbols, is
generated before reaching the end of the vector. In this latter case, the extra dimension values are simply ignored and considered introns that may be
switched on in subsequent iterations. Although the vectors were bounded in
length in earlier studies not all elements were necessarily used to construct
a program during the mapping process, and as such the programs generated
were variable in size.

3.4.1 Variable-Length Particle Strategies

Four different approaches to a variable-length particle swarm algorithm were
investigated in this study.

Strategy I

Each particle in the swarm is compared to the global best particle (gbest) to
determine if there is a difference between the length of the particle's vector
and the length of the gbest vector. If there is no difference between the vector
sizes then a length update is not required and the algorithm simply moves on
and compares the next particle to gbest. However, when there is a difference
between the vector lengths, the particle is either extended or truncated. If the
current particles, p_i vector length is shorter than the length of gbest, elements
are added to the particle's vector extending it so that it is now equivalent in
length to that of gbest. The particle's new elements contain values which are
copied directly from gbest. For example, if gbest is a vector containing fifty
elements and the current particle has been extended from forty five to fifty
elements then the values contained in the last 5 elements (46-50) of gbest are
copied into the five new elements of the current particle. If the particle has a
greater number of elements than the gbest particle, then the extra elements are
simply truncated so that both gbest and the current particle have equivalent
vector lengths.

Strategy II

This strategy is similar to the first strategy, the only difference is the method in which the new elements are copied. In the first strategy, when the current particle, p_i is extended the particle's new elements are populated by values which are copied directly from gbest. In Strategy II, values are not copied from gbest instead random numbers are generated in the range [cmin, cmax] and these values are copied into each of p_i's new elements.

Strategy III

The third strategy involves the use of probabilities. Given a specified probability, the length of the particle is either increased or decreased. A maximum of one elements can only be changed at a time i.e. either an element is added or removed from the current particle, p_i. If p_i is longer than gbest then the last element of p_i is discarded. If p_i is shorter than gbest then p_i is increased by adding an extra element to its vector. In this situation the new element takes the value of a random number in the range [cmin, cmax].

Strategy IV

The fourth strategy involves the generation of a random number to determine the number of elements that will be added to or removed from the current particle, p_i. If the length of p_i is shorter than the length of gbest the difference, dif between the length of gbest and the length of p_i is calculated. Then a random integer is generated in the range $[0, dif]$. The result of this calculation is then used to determine how many elements will be truncated from p_i. A similar strategy is applied when the length of p_i is smaller than the length of gbest. However, in this case the random number generated is used to determine the number of elements that p_i will be extended by. After p_i is extended, each of these extended elements are then populated with random numbers generated in the range [cmin, cmax].

A strategy is not applied every time it was possible to modify the current particle (p_i), instead applying a strategy is determined by the outcome of a certain probability function i.e. the outcome of this function is used to determine if a strategy is to be applied to p_i. In our current implementation, a probability of 0.5 was selected. Therefore 50% of the time a length-modifying strategy is applied and 50% of the time the length of p_i is not modified.

For each particle in the swarm, a random number in the range [1,100] is generated, which determines its initial length in terms of the number of codons.

3.5 Proof of Concept Experiments and Results

A diverse selection of benchmark programs from the literature on Genetic Programming are tackled using Grammatical Swarm to demonstrate proof of concept for the variable-length GS methodology. The parameters adopted across the following experiments are $c_1 = c_2 = 1.0$, $wmax = 0.9$, $wmin = 0.4$, $cmin = 0$ (minimum value a coordinate may take), $cmax = 255$ (maximum value a coordinate may take). In addition, a swarm size of 30 running for 1000 iterations is used, and 100 independent runs are performed for each experimental setup with average results being reported.

The same problems are also tackled with GS's fixed-length counterpart (using 100 dimensions) and GE in order to determine how well the variable-length GS algorithm is performing at program generation in relation to the more traditional variable-length Genetic Algorithm search engine of standard GE. In an attempt to achieve a relatively fair comparison of results given the differences between the search engines of Grammatical Swarm and Grammatical Evolution, we have restricted each algorithm in the number of individuals they process. Grammatical Swarm running for 1000 iterations with a swarm size of 30 processes 30,000 individuals, therefore, a standard population size of 500 running for 60 generations is adopted for Grammatical Evolution. The remaining parameters for Grammatical Evolution are roulette selection, steady state replacement, one-point crossover with probability of 0.9, and a bit mutation with probability of 0.01.

3.5.1 Santa Fe Ant trail

The Santa Fe ant trail is a standard problem in the area of GP and can be considered a deceptive planning problem with many local and global optima [12]. The objective is to find a computer program to control an artificial ant so that it can find all 89 pieces of food located on a non-continuous trail within a specified number of time steps, the trail being located on a 32 by 32 toroidal grid. The ant can only turn left, right, move one square forward, and may also look ahead one square in the direction it is facing to determine if that square contains a piece of food. All actions, with the exception of looking ahead for food, take one time step to execute. The ant starts in the top left-hand corner of the grid facing the first piece of food on the trail. The grammar used in this problem is different to the ones used later for symbolic regression and the multiplexer problem in that we wish to produce a multi-line function in this case, as opposed to a single line expression. The grammar for the Santa Fe ant trail problem is given below.

```
<code> ::= <line> | <code> <line>
<line> ::= <condition> | <op>
<condition> ::= if(food_ahead()) { <line> } else { <line> }
<op> ::= left(); | right(); | move();
```

3.5.2 Quartic Symbolic Regression

The target function is $f(a) = a + a^2 + a^3 + a^4$, and 100 randomly generated input-output vectors are created for each call to the target function, with values for the input variable drawn from the range [0,1]. The fitness for this problem is given by the reciprocal of the sum, taken over the 100 fitness cases, of the absolute error between the evolved and target functions. The grammar adopted for this problem is as follows:

```
<expr> ::= <expr> <op> <expr> | <var>
<op> ::=  + | - | * | /
<var> ::= a
```

3.5.3 Three Multiplexer

An instance of a multiplexer problem is tackled in order to further verify that it is possible to generate programs using Grammatical Swarm. The aim with this problem is to discover a boolean expression that behaves as a 3 Multiplexer. There are 8 fitness cases for this instance, representing all possible input-output pairs. Fitness is the number of input cases for which the evolved expression returns the correct output. The grammar adopted for this problem is as follows:

```
<mult> ::= guess = <bexpr> ;
<bexpr> ::= ( <bexpr> <bilop> <bexpr> )
           | <ulop> ( <bexpr> )
           | <input>
<bilop> ::= and | or
<ulop> ::= not
<input> ::= input0 | input1 | input2
```

3.5.4 Mastermind

In this problem the code breaker attempts to guess the correct combination of colored pins in a solution. When an evolved solution to this problem (i.e. a combination of pins) is to be evaluated, it receives one point for each pin that has the correct color, regardless of its position. If all pins are in the correct order then an additional point is awarded to that solution. This means that ordering information is only presented when the correct order has been found for the whole string of pins.

A solution therefore, is in a local optimum if it has all the correct colors, but in the wrong positions. The difficulty of this problem is controlled by the number of pins and the number of colors in the target combination. The instance tackled here uses 4 colors and 8 pins with the following target values 3 2 1 3 1 3 2 0.

The grammar adopted is as follows.

```
<pin> ::= <pin> <pin> | 0 | 1 | 2 | 3
```

3.5.5 Results

Results averaged over 100 runs showing the best fitness, and the cumulative frequency of success for the four variable length grammatical swarm (VGS) variants are presented in Fig. 3.2, Fig. 3.3, Fig. 3.4 and Fig. 3.5.

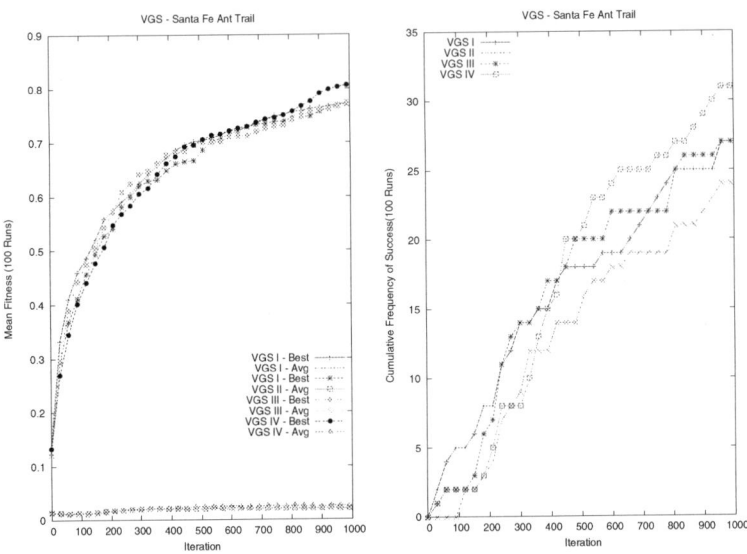

Fig. 3.2. Plot of the mean fitness on the Santa Fe Ant trail problem instance (left), and the cumulative frequency of success (right).

Table 3.1, Table 3.2, Table 3.3 and Table 3.4 outline a comparison of the results across the four variable-length particle swarm strategies analysed in this study. While there is no clear winner across all four problems strategies, III and IV were the most successful overall, with strategy IV producing best performance on the Santa Fe ant and Multiplexer problems, while strategy III had the better performance on the Symbolic Regression and Mastermind instances. It is interesting to note that the mean length of the gbest particle never grows beyond 65 codons at the last iteration across all four problems, demonstrating that bloat does not appear to have impacted on these results.

3.5.6 Summary

Table 3.5 provides a summary and comparison of the performance of the fixed and variable-length forms of GS and GE on each of the problem domains tackled. The best variable-length strategy outperforms GE on the Mastermind instance and has a similar performance to the fixed-length form of GS. On the

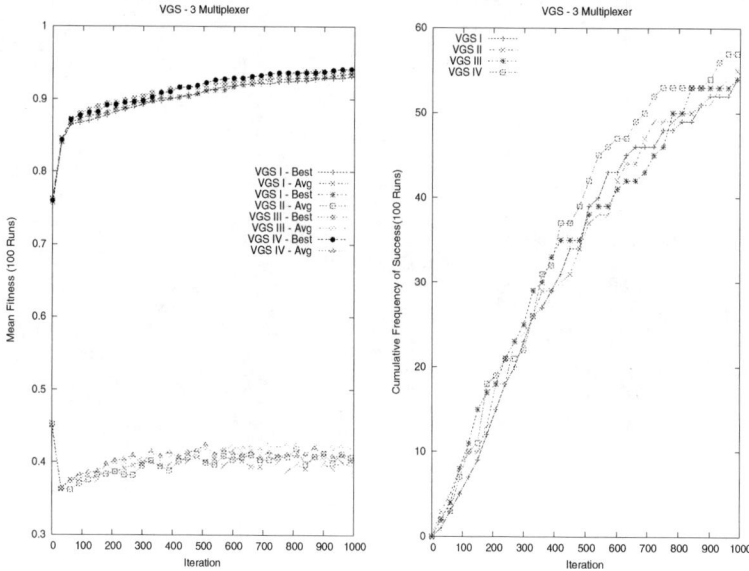

Fig. 3.3. Plot of the mean fitness on the 3 multiplexer problem instance (left), and the cumulative frequency of success (right).

Table 3.1. A comparison of the results obtained for the four different variable-length Particle Swarm Algorithm strategies on the Santa Fe Ant trail.

	Mean Best	Successful	Mean gbest
	Fitness	Runs	Codon Length
Strategy			
I	.77	27	50
II	.76	24	51
III	.78	27	51
IV	.8	31	61

Table 3.2. A comparison of the results obtained for the four different variable-length Particle Swarm Algorithm strategies on the Multiplexer problem instance.

	Mean Best	Successful	Mean gbest
	Fitness	Runs	Codon Length
Strategy			
I	.93	54	49
II	.94	55	52
III	.94	54	57
IV	.94	57	53

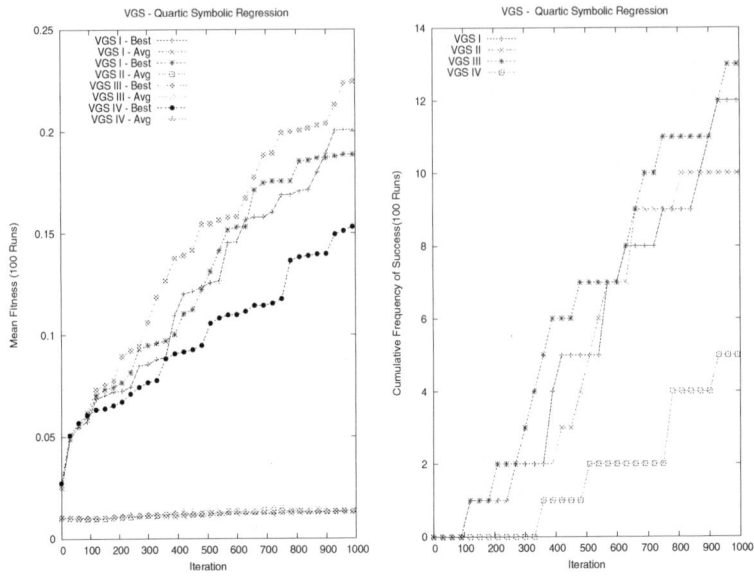

Fig. 3.4. Plot of the mean fitness on the Quartic Symbolic Regression problem instance (left), and the cumulative frequency of success (right).

Table 3.3. A comparison of the results obtained for the four different variable-length Particle Swarm Algorithm strategies on the quartic symbolic regression problem instance.

	Mean Best Fitness	Successful Runs	Mean gbest Codon Length
Strategy			
I	.2	12	45
II	.19	10	49
III	.23	13	55
IV	.15	5	54

Table 3.4. A comparison of the results obtained for the four different variable-length Particle Swarm Algorithm strategies on the Mastermind problem.

	Mean Best Fitness	Successful Runs	Mean gbest Codon Length
Strategy			
I	.89	10	61
II	.9	12	57
III	.9	14	65
IV	.9	12	60

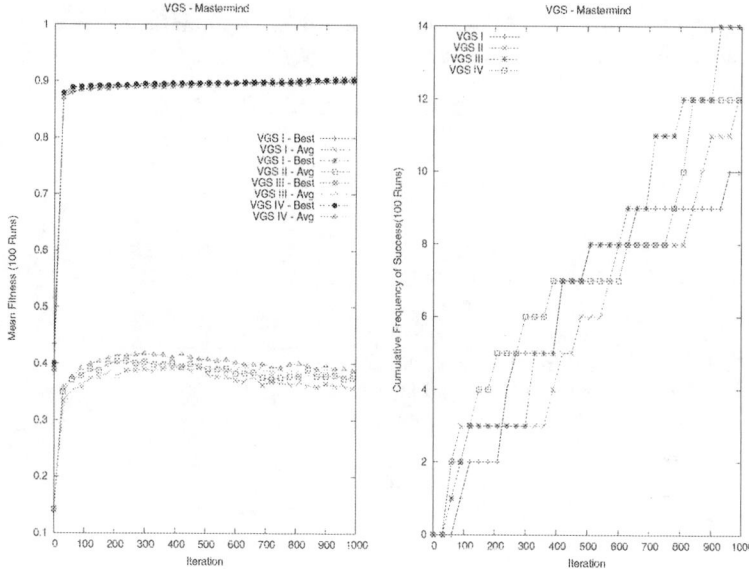

Fig. 3.5. Plot of the mean fitness on the Mastermind problem instance (left), and the cumulative frequency of success (right).

other three problems the fixed-length form of GS outperforms variable-length GS in terms of the number of successful runs finding the target solution. On both the Santa Fe ant and Symbolic Regression problems, GE outperforms GS. The key finding is that the results demonstrate proof of concept that a variable-length particle swarm algorithm can successfully generate solutions to problems of interest. In this initial study, we have not attempted parameter optimization for the various variable-length strategies and this may lead to further improvements of the variable-length particle swarm algorithm. Given the relative simplicity of the Swarm algorithm, the small population sizes involved, and the complete absence of a crossover operator synonymous with program evolution in GP, it is impressive that solutions to each of the benchmark problems have been obtained. Based on the findings in this study there is no clear winner between the bounded and variable-length forms of GS, and as such the recommendation at present would be to adopt the simpler bounded GS, although future investigations may find in the variable-length algorithm's favour.

3.6 Conclusions and Future Work

This study demonstrates the feasibility of successfully generating computer programs using a variable-length form of Grammatical Swarm, and demon-

Table 3.5. A comparison of the results obtained for Grammatical Swarm and Grammatical Evolution across all the problems analyzed.

	Successful Runs
Santa Fe ant	
GS (variable)	31
GS (bounded)	38
GE	58
Multiplexer	
GS (variable)	57
GS (bounded)	87
GE	56
Symbolic Regression	
GS (variable)	13
GS (bounded)	28
GE	85
Mastermind	
GS (variable)	14
GS (bounded)	13
GE	10

strates its application to a diverse set of benchmark program-generation problems. A performance comparison to Grammatical Evolution has shown that Grammatical Swarm is on a par with Grammatical Evolution, and is capable of generating solutions with much smaller populations, with a fixed-length vector representation, an absence of any crossover, and no concept of selection or replacement. A performance comparison of the variable-length and fixed-length forms of Grammatical Swarm reveal that the simpler fixed-length version is superior for the experimental setups and problems examined here.

The results presented are very encouraging for future development of the relatively simple Grammatical Swarm algorithm, and other potential Social or Swarm Programming variants.

References

1. Banzhaf, W., Nordin, P., Keller, R.E. and Francone, F.D. (1998). *Genetic Programming – An Introduction; On the Automatic Evolution of Computer Programs and its Applications*. Morgan Kaufmann.
2. Bonabeau, E., Dorigo, M. and Theraulaz, G. (1999). *Swarm Intelligence: From natural to artificial systems*, Oxford: Oxford University Press.
3. Brabazon, A. and O'Neill, M. 2006. *Biologically Inspired Algorithms for Financial Modelling*. Springer.
4. Cleary, R. and O'Neill, M. 2005. An Attribute Grammar Decoder for the 01 MultiConstrained Knapsack Problem. In LNCS 3448 *Pr oc. of Evolutionary Computation in Combinatorial Optimization EvoCOP 2005*, pp.34-45, Lausanne, Switzerland. Springer.

5. Hemberg, M. and O'Reilly, U-M. 2002. GENR8 - Using Grammatical Evolution In A Surface Design Tool. In *Proc. of the First Gra mmatical Evolution Workshop GEWS2002*, pp.120-123. New York City, New York, US. ISGEC.
6. Kennedy, J., Eberhart, R. and Shi, Y. (2001). *Swarm Intelligence*, San Mateo, California: Morgan Kauffman.
7. Kennedy, J. and Eberhart, R. (1995). Particle swarm optimization, *Proceedings of the IEEE International Conference on Neural Networks*, December 1995, pp.1942-1948.
8. Koza, J.R. (1992). *Genetic Programming*. MIT Press.
9. Koza, J.R. (1994). *Genetic Programming II: Automatic Discovery of Reusable Programs*. MIT Press.
10. Koza, J.R., Andre, D., Bennett III and F.H., Keane, M. (1999). *Genetic Programming 3: Darwinian Invention and Problem So lving*. Morgan Kaufmann.
11. Koza, J.R., Keane, M., Streeter, M.J., Mydlowec, W., Yu, J., Lanza, G. (2003). *Genetic Programming IV: Routine Human-Co mpetitive Machine Intelligence*. Kluwer Academic Publishers.
12. Langdon, W.B. and Poli, R. (1998). Why Ants are Hard. In *Genetic Programming 1998: Proceedings of the Th ird Annual Conference*, University of Wisconsin, Madison, Wisconsin, USA, pp. 193-201, Morgan Kaufmann.
13. Moore, J.H. and Hahn, L.W. (2004). Systems Biology Modeling in Human Genetics Using Petri Nets and Grammatical Evolution . In LNCS 3102 *Proc. of the Genetic and Evolutionary Computation Conference GECCO 2004*, Seattle, WA, USA, pp.392-401. Springer.
14. O'Neill, M. and Brabazon, A. (2005). Recent Adventures in Grammatical Evolution. In *Computer Methods and Systems CMS'05*, Krakow, Poland, pp.245-252. Oprogramowanie Naukowo-Techniczne.
15. O'Neill, M. and Brabazon, A. (2004). Grammatical Swarm. In LNCS 3102 *Proc. of the Genetic and Evolutionary Computation Conferen ce GECCO 2004*, Seattle, WA, USA, pp.163-174. Springer.
16. O'Neill, M., Adley, C. and Brabazon, A. (2005). A Grammatical Evolution Approach to Eukaryotic Promoter Recognition. In *Proc. of Bioinformatics IN-FORM 2005*, Dublin City University, Dublin, Ireland.
17. O'Neill, M., Brabazon, A. and Adley, C. (2004). The automatic generation of programs for Classification using Grammatical Swarm. In *Proc. of the Congress on Evolutionary Computation CEC 2004*, Portland, OR, USA, pp.104-110. IEEE.
18. O'Neill, M. and Ryan, C. (2003). *Grammatical Evolution: Evolutionary Automatic Programming in an Arbitrary Language*. Kluwer Academic Publishers.
19. O'Neill, M. (2001). *Automatic Programming in an Arbitrary Language: Evolving Programs in Grammatical Evolution*. PhD thesis, University of Limerick, 2001.
20. O'Neill, M. and Ryan, C. (2001). Grammatical Evolution, *IEEE Trans. Evolutionary Computation*. 2001.
21. O'Neill, M., Ryan, C., Keijzer M. and Cattolico M. (2003). Crossover in Grammatical Evolution. *Genetic Programming and E volvable Machines*, Vol. 4 No. 1. Kluwer Academic Publishers, 2003.
22. Ryan, C., Collins, J.J. and O'Neill, M. (1998). Grammatical Evolution: Evolving Programs for an Arbitrary Language. *Proc. of the First European Workshop on GP*, 83-95, Springer-Verlag.

23. Silva, A., Neves, A. and Costa, E. (2002). An Empirical Comparison of Particle Swarm and Predator Prey Optimisation. In *LN AI 2464, Artificial Intelligence and Cognitive Science, the 13th Irish Conference AICS 2002*, pp. 103-110, Limerick, Ireland, Springer.

4

SWARMs of Self-Organizing Polymorphic Agents

Derek Messie[1] and Jae C. Oh[2]

[1] Department of Electrical Engineering and Computer Science, Syracuse University, Syracuse, NY USA.
dsmessie@syr.edu, http://messie.syr.edu
[2] Department of Electrical Engineering and Computer Science, Syracuse University, Syracuse, NY USA.
jcoh@ecs.syr.edu, http://web.syr.edu/j̃coh

The field of Swarm Intelligence is increasingly being seen as providing a framework for solving a wide range of large-scale, distributed, complex problems. Of particular interest are architectures and methodologies that address organization and coordination of a large number of relatively simple agents distributed across the system in a way that encourages some desirable global emergent behavior. This chapter describes a SWARM simulation of a distributed approach to fault mitigation within a large-scale data acquisition system for BTeV, a particle accelerator-based High Energy Physics experiment currently under development at Fermi National Accelerator Laboratory. Incoming data is expected to arrive at a rate of over 1 terabyte every second, distributed across 2500 digital signal processors. Simulation results show how lightweight polymorphic agents embedded within the individual processors use game theory to adapt roles based on the changing needs of the environment. SWARM architecture and implementation methodologies are detailed.

4.1 Introduction

In the field of Swarm Intelligence, a lot of attention has been focused lately on developing large-scale distributed systems that are capable of coordinating individual actors in a system competing for resources such as bandwidth, computing power, and data. Agent methodologies that exhibit self-* (self-organizing, self-managing, self-optimizing, self-protecting) attributes are of particular value [5, 12]. SWARM (http://www.swarm.org), is a software development kit that allows for large-scale simulations of complex multi-agent systems. The SWARM experiments conducted demonstrate *polymorphic* self-*

D. Messie and J.C. Oh: *SWARMs of Self-Organizing Polymorphic Agents*, Studies in Computational Intelligence (SCI) **26**, 75–90 (2006)
www.springerlink.com

agents that are capable of multiple roles as directed strictly by the environment. These agents evolve an optimum core set of roles for which they are responsible, while still possessing the ability to take on alternate roles as environmental demands change.

SWARM is used to simulate the RTES/BTeV environment, a data acquisition system for a particle accelerator-based High Energy Physics experiment currently under development at Fermi National Accelerator Laboratory. Multiple layers of polymorphic, very lightweight agents (VLAs) are embedded within 2500 Digital Signal Processors (DSPs) to handle fault mitigation across the system.

This chapter is divided into six sections. First, some background on polymorphism and stigmergy, along with the RTES/BTeV environment itself is provided. A description of VLAs embedded within Level 1 of the RTES/BTeV environment is provided, followed by an explanation of current challenges and other motivating factors. Details of the SWARM development kit components and configuration used in the experiments are provided Section 4.3, as well as an introduction to polymorphic self-* agents design. Results of the SWARM simulation are evaluated in Section 4.5, followed by lessons learned, next steps, and conclusions.

4.2 Background and Motivation

4.2.1 Polymorphism and Stigmergy

Polymorphism and stigmergy are founded in biology and the study of self-organization within social insects. The term *polymorphism* is often used in describing characteristics of ants and other social biological systems, and is defined as the occurrence of different forms, stages, or types in individual organisms, or organisms of the same species, independent of sexual variations [19, 11]. Within individual colonies consisting of ants with the same genetic wiring, two or more castes belonging to the same sex can be found. The function or role that any one ant takes on is dictated by cues from the environment [18].

The agents described in this chapter adhere to this definition of polymorphism in that they are genetically identical, yet each evolve distinct roles that they play as demanded of them through changes in the environment.

The concept of polymorphic agents presented in this chapter is different from other definitions of polymorphism that have surfaced in computer science. In object-oriented programming, polymorphism is usually associated with the ability of objects to override inherited class method implementations [8]. The term has also arisen in other subareas of computer science, including some agent designs [1], but generally describes a templating based system or similar variation of the object-oriented model.

(a) Large termite mound

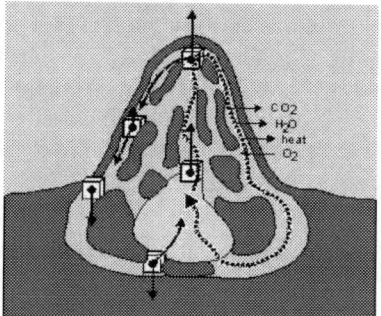
(b) Mound as respiratory device

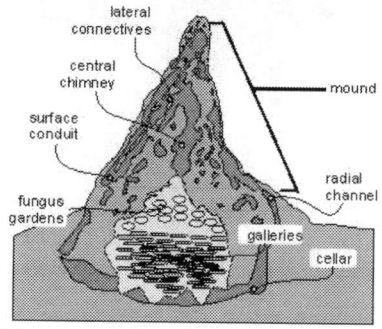
(c) Air shafts to underground spaces

Fig. 4.1. (a) Large termite mound commonly found in subsaharan Africa. (b) The mounds act as respiratory devices, built from the surrounding soil by the termites in a colony. The mound powers ventilation of the subterranean nest by capturing energy from the wind. (c) Air shafts lead to underground spaces where larvae, food, and fungus is stored.

Stigmergy was introduced by biologist Pierre-Paul Grasse to describe indirect communication that takes place between individuals in social insect societies [6]. The theory explains how organization and coordination of the building of termite nests is mainly controlled by the nest itself, and not the individual termite workers involved. It views the process of emergent cooperation as a result of participants altering the environment and reacting to the environment as they pass through it. The canonical example of stigmergy is ants leaving pheromones in ways that help them find the shortest, safest distance to food or to build nests.

A stigmergic approach to fault mitigation is introduced in this chapter. Individual agents communicate indirectly through errors that they find (or do not find) in the environment. This indirect communication is manifested through actions that each agent takes as cued by the environment. Results show how the local actions of polymorphic agents within the system allow self-* global behavior to emerge.

Polymorphism and stigmergy are two of the core elements that produce complex organization and coordination within social insect societies. Although individual participants rely on local information only to act on a small set of basic rules, very complex behavior and structure can emerge. Fig. 4.1(a) shows a large complex termite mound commonly found in subsaharan Africa. As detailed in 4.1(b), the mounds are respiratory devices, built from the surrounding soil by the termites in a colony. The mound powers ventilation of the subterranean nest by capturing energy from wind. They are organs of colony physiology, shaped to accommodate and regulate the exchanges of respiratory gases between the nest and atmosphere. As labeled in Fig. 4.1(c), the air shafts within the mounds also lead underground to the cellar, where larvae, food, and fungus is stored.

4.2.2 RTES/BTeV

BTeV is a proposed particle accelerator-based HEP experiment currently under development at Fermi National Accelerator Laboratory in Chicago, IL. The goal is to study charge-parity violation, mixing, and rare decays of particles known as beauty and charm hadrons, in order to learn more about matter-antimatter asymmetries that exist in the universe today [10].

An aerial view of the Fermilab Tevatron is shown in Fig. 4.2. The experiment uses approximately 30 planar silicon pixel detectors that are connected to specialized field-programmable gate arrays (FPGAs). The FPGAs are connected to approximately 2500 digital signal processors (DSPs) that filter incoming data at the extremely high rate of approximately 1.5 Terabytes per second from a total of 20×10^6 data channels. A three tier hierarchical trigger architecture will be used to handle this high rate [10]. An overview of the BTeV triggering and data acquisition system is shown in Fig. 4.3, including a magnified view of the L1 Vertex Trigger responsible for Level 1 filtering consisting of 2500 Worker nodes (2000 Track Farms and 500 Vertex Farms).

There are many Worker level tasks that the Farmlet VLA (FVLA) is responsible for monitoring. A traditional hierarchical approach would assign one (or more) distinct DSPs the role of the FVLA, with the responsibility of monitoring the state of other Worker DSPs on the node [3]. However, this leaves the system with only very few possible points of failure before critical tasks are left unattended.

Another approach would be to assign a single redundant DSP (or more) to each and every Worker DSP, to act as the FVLA [7]. However, since 2500 Worker DSPs are projected, this would prove very expensive and may still not fully protect all DSPs given even a low number of system failures.

The events that pass the full set of physics algorithm filters occur very infrequently, and the cost of operating this environment is high. The extremely large streams of data resulting from the BTeV environment must be processed real-time with highly resilient adaptive fault tolerant systems.

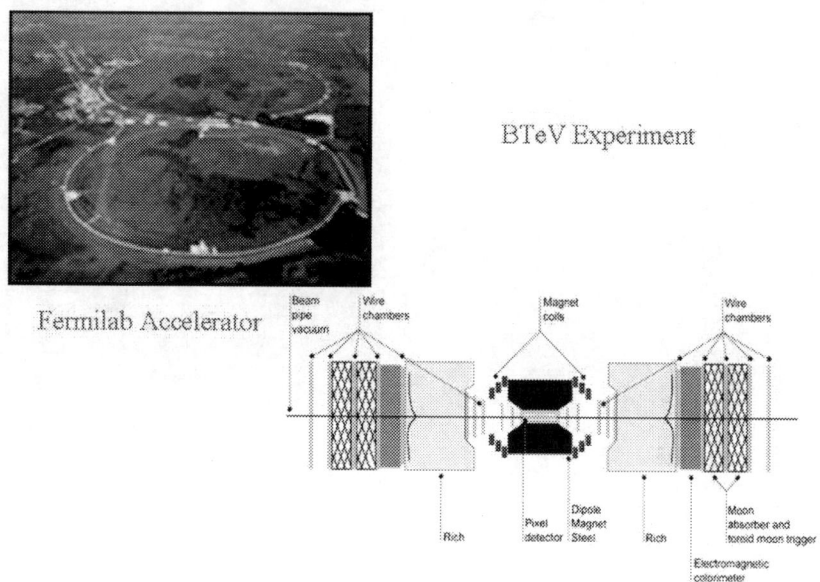

Fig. 4.2. Aerial view of the Fermilab Tevatron, the world's highest-energy particle collider. Beams of protons and antiprotons are collided to examine the basic building blocks of matter.

4.2.3 Very Lightweight Agents (VLAs)

Multiple levels of very lightweight agents (VLAs) [16] are one of the primary components responsible for fault mitigation across the BTeV data acquisition system.

The primary objective of the VLA is to provide the BTeV environment with a lightweight, adaptive layer of fault mitigation. One of the latest phases of work at Syracuse University has involved implementing embedded proactive and reactive rules to handle specific system failure scenarios.

VLAs have been implemented in two separate scaled prototypes of the RTES/BTeV environment. The first was presented at the SuperComputing 2003 (SC2003) conference [13], and the other at the 11th IEEE Real-Time and Embedded Technology and Applications Symposium (RTAS2005). Reactive and proactive VLA rules were integrated within these Level 1 and Level 2 prototypes and served a primary role in demonstrating the embedded fault tolerant capabilities of the system.

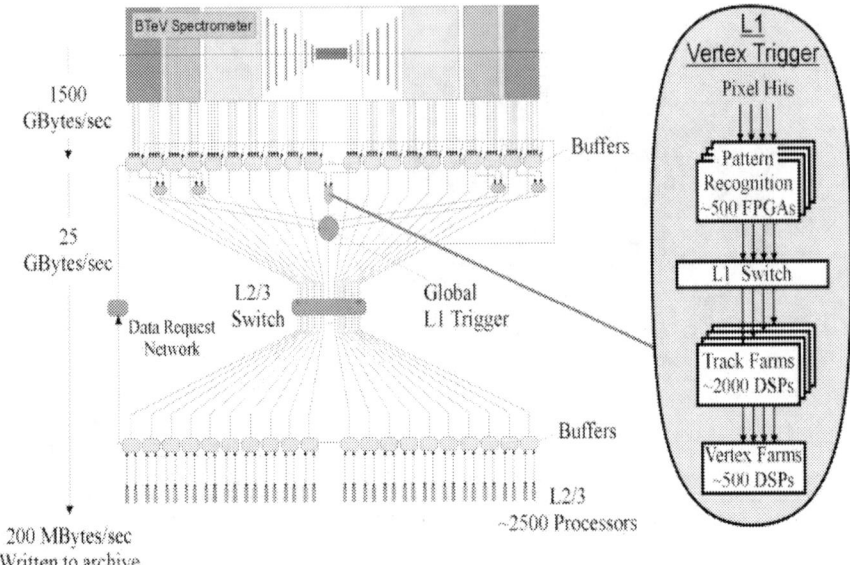

Fig. 4.3. The BTeV triggering and data acquisition system showing (left side) detector, buffer memories, L1, L2, L3 clusters and their interconnects and (right side) a magnified figure of the L1 Vertex trigger.

4.2.4 Challenges

While these prototypes were effective for demonstrating the real-time fault mitigation capabilities of VLAs on limited hardware utilizing 16 DSPs, one of the major challenges is to find out how the behavior of the various levels of VLAs will scale when implemented across the 2500 DSPs projected for BTeV [9]. In particular, how frequently should these monitoring tasks be performed to optimize available processing time, and what affect does this have on other components and the overall behavior of a large-scale real-time embedded system such as BTeV.

Given the number of components and countless fault scenarios involved, it is infeasible to design an 'expert system' that applies mitigative actions triggered from a central processing unit acting on rules capturing every possible system state. Instead, Section 4.4 describes a distributed approach that uses self-organizing VLAs to accomplish fault mitigation within the large-scale real-time RTES/BTeV environment.

4.3 SWARM Simulation of RTES/BTeV

4.3.1 Overview

The goal of the implementation is to evaluate the behavior of the VLAs and other components using a high volume of rules across 2500 DSPs. This requires a simulation environment that will allow abstract representation of some of the complex integration within BTeV.

SWARM (http://www.swarm.org), distributed under the GNU General Public License, is software available as a Java or Objective-C development kit that allows for the multi-agent simulation of complex systems [2, 4]. It consists of a set of libraries that facilitate implementation of agent-based models. SWARM has previously been used by the RTES team in simulations that model the RTES/BTeV environment [15].

The basic architecture of SWARM provides for the simulation of collections of concurrently interacting agents. It provides an environment that can model various components of the BTeV system, assigning dynamic states to each agent, which can then be altered in time steps following various user-specified rules. Both proactive and reactive rules are triggered after the current state of a given agent (or component) is evaluated against the state of other connected agents (or components).

4.3.2 SWARM Development Kit

The SWARM development kit consists of a core set of libraries, each of which are described completely in the SWARM Documentation Set [3]. The top application layer (known as the 'observer swarm') creates screen displays, along with all lower levels. The next level down ('model swarm') then creates individual agents, schedules their activities, collects information about them, and relays that information when requested to do so by the 'observer swarm'.

The SWARM libraries also provide numerous additional objects for designing the agent-based model, including tools that facilitate the management of memory, the maintenance of lists, and the scheduling of actions.

The state of each worker is represented with distinct error flags that represent the unique error code(s) that each worker is experiencing at any point in time. Each VLA has an error log which tracks the error messages that it receives at each time step.

Since scalability issues are a primary concern, the results must detect the degree to which any exhibited behavior is tied to specific system configurations. The SWARM model includes dynamic variables that can be modified to reflect various hardware layout configurations, including the number of PAs per Farmlet, the number of Farmlets, the number of Farmlets per Region, the number of Regions, and the frequency rate of individual failure scenarios.

---

[3] SWARM Documentation Set is found at http://www.swarm.org/swarmdocs/set/set.html

4.3.3 Polymorphic Agents

Some initial supporting terminology from the field of multi-agent systems should first be presented before a definition for polymorphic agents is shown. To begin with, finding a single universally accepted definition of an *agent* has been as difficult for the agent-based computing community, as defining *intelligence* has been for the mainstream AI community [20]. However, a few of the core properties of agents that have been widely accepted include :

- *autonomy*: agents operate without the direct intervention of humans or others, and have some kind of control over their actions and internal state;
- *social ability*: agents interact with other agents;
- *reactivity*: agents respond to changes in their environment;
- *pro-activeness*: agents exhibit goal-directed behavior.

Likewise, robust definitions of *environments* in multi-agent systems can be even more challenging to find than those for *agents*. In very general terms, environments 'provide the conditions under which an entity (agent or object) exists' [17].

In multi-agent systems, roles are viewed as 'an abstract representation of an agents function, service, or identification within a group' [14]. As described earlier, *polymorphism* is defined in biology as:

'the occurrence of different forms, stages, or types in individual organisms or in organisms of the same species, independent of sexual variations' [19, 11].

For the definition of *polymorphic agents* in multi-agent systems proposed in this chapter, *organisms* are replaced by *agents*, and agent *roles* (function, service, or identification) are used in place of *forms*, *stages*, and *types* of organisms. The environment in this new definition in many ways remains unchanged.

This leads to the following working definition for *polymorphic agents* within the field of multi-agent systems:

Polymorphic Agents -

'Individual agents within groups of similar or identical agents that are capable of adapting roles based on their perceived environment.'

SWARM is used to evaluate a stigmergic multi-agent systems approach using polymorphic agents to address the weaknesses inherent in traditional hierarchical fault mitigation designs. In this model, rather than hard-wiring the assignment of FVLA roles to specialized FVLAs on dedicated DSPs, Worker VLAs are made polymorphic so that *every* VLA is equipped to play the role of FVLA for *any* DSP on the same node.

Since the FVLA is responsible for a wide range of monitoring tasks, this means that we must build the capability of performing each task into every Worker Level VLA. The classic problem this presents in traditional hierarchical approaches is how to process all of the data necessary for all of these tasks in time for a useful response [20]. However, since these agents are polymorphic and evolve roles gradually over time, there is only a small set of tasks for which each agent is responsible for at any given point in time.

Stigmergy is used to determine which set of tasks any given VLA performs. Errors found (or not found) in the environment by an individual VLA increase (decrease) the sensitivity of that VLA to that particular type of error. Agents start out by monitoring each type of error at a fixed rate. Then, based entirely on what is encountered in the environment, each develops a core set of roles for which it takes responsibility.

Game theory is used to facilitate self-organization within the SWARM of agents. Each DSP calculates a utility value to determine locally precisely when the PA or VLA should have control of the DSP. The utility value used in this implementation is based on the processing cost of performing FVLA monitoring tasks, and the estimated benefit of performing the tasks. The utility of a single DSP (i) performing an FVLA monitoring task on another DSP (j) is :

$$u_{ij} = -c_{ij} + b_{ij} * p(t), \tag{4.1}$$

where c_{ij} is the cost of DSP i performing the FVLA monitoring task on DSP j, b_{ij} is the estimated benefit received by DSP i finding an error on DSP j, and $p(t) = 2 * ((1/(1+e^{-dt})) - .5)$, an adjusted sigmoid value for the amount of time elapsed (t) since DSP i last performed the monitoring task on DSP j.

It is important to note here that the value assigned to d in the sigmoid value for the amount of time elapsed, determines the steepness of the sigmoid function, and hence the sensitivity of the agent to a given error. In other words, the higher the value of d, the higher the adjusted sigmoid value of t, and the higher the sensitivity (the frequency of checks) of the VLA to a particular error.

This is where the polymorphic behavior of the VLA is demonstrated. Any time that an individual VLA finds a specific error while performing FVLA monitoring tasks, the d value for that error on that particular node is increased. Any time that an individual VLA performs a monitoring task and does *not* find an error, the d value is slightly decreased. A high value for d means that FVLA tasks are performed more frequently (high sensitivity), whereas a low value for d means they are performed less often (low sensitivity).

A low d value means low steepness (sensitivity), which means the utility of performing FVLA checks increases slowly over time. A high d value on the other hand, means high steepness (sensitivity), and utility increases sharply over time.

utility

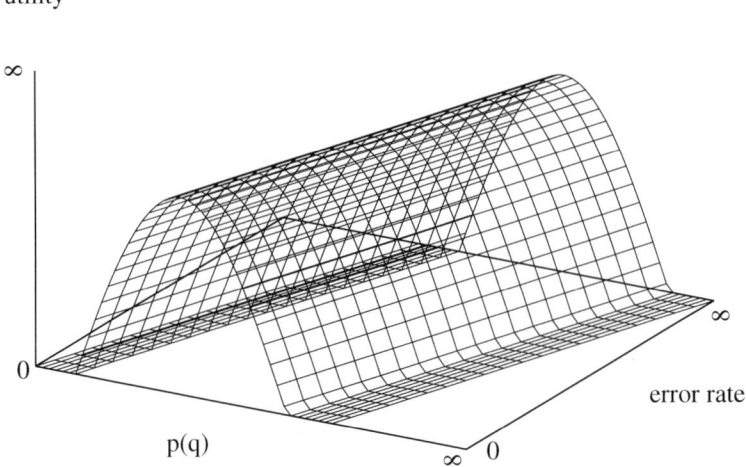

Fig. 4.4. Utility based on the sigmoid value for the frequency of FVLA checks (p(q)) on a given DSP. The optimum value for the frequency of checks changes as the error rate fluctuates. A higher error rate calls for a higher rate of checking to reach optimum utility, whereas a lower error rate requires less frequent checks.

The *total* utility of a single DSP (i) performing FVLA monitoring tasks across all n DSPs is thus :

$$U_i = \sum_{j=1}^{n}((-c_{ij} + b_{ij} * p(t)) * \alpha) \quad , i \neq j, \tag{4.2}$$

where c_{ij} is the cost of DSP i performing the FVLA monitoring task on DSP j, b_{ij} is the estimated benefit received by DSP i finding an error on DSP j, $p(t) = 2 * ((1/(1 + e^{-dt})) - .5)$, an adjusted sigmoid value for the amount of time elapsed (t) since DSP i last performed the monitoring task on DSP j, and α is a flag set to 1 if the monitoring task is performed, and 0 if it is not.

Next, a utility function is defined that takes into account the quantity of FVLA monitoring tasks performed by a single DSP (i) on another DSP (j) over a fixed time period. The following equation calculates the utility of a DSP performing FVLA monitoring tasks based on the frequency of checks over a given time period, along with the cost of performing the checks :

$$u_{ij} = -c_{ij} * q_{ij} + b_{ij} * p(q_{ij}), \tag{4.3}$$

where c_{ij} is the cost of DSP i performing the FVLA monitoring task on DSP j, q_{ij} is the number of times DSP i performed the monitoring task on DSP j over a fixed time interval, b_{ij} is the estimated benefit received by DSP i finding an error on DSP j, and $p(q_{ij}) = 2 * ((1/(1 + e^{-dq_{ij}})) - .5)$, an adjusted sigmoid value for the number of times DSP i performed the monitoring task on DSP j over a fixed time period.

This utility value allows us to calculate the optimum number of FVLA checks that should be performed by a given DSP over a set time period. As described earlier, if checks are performed too frequently or not freqently enough, then DSP clock cycles are wasted (utility decreases). Fig. 4.4 demonstrates this by showing the effect that the frequency of checks has on utility. It also shows how the optimum value for the frequency of checks is affected as the error rate fluctuates.

The following equation is used to represent the total utility of a single DSP performing FVLA monitoring tasks across all n DSPs.

$$U_i = \sum_{j=1}^{n} (-c_{ij} * q_{ij} + b_{ij} * p(q_{ij})) \quad , i \neq j, \quad (4.4)$$

where c_{ij} is the cost of DSP i performing the FVLA monitoring task on DSP j, q_{ij} is the number of times DSP i performed the monitoring task on DSP j over a fixed time interval, b_{ij} is the estimated benefit received by DSP i finding an error on DSP j, and $p(q_{ij}) = 2 * ((1/(1 + e^{-dq_{ij}})) - .5)$, an adjusted sigmoid value for the number of times DSP i performed the monitoring task on DSP j over a fixed time period.

4.4 Results

4.4.1 Summary

SWARM simulates Farmlet data buffer queues that are populated at a rate consistent with the actual incoming physics crossing data. Each DSP within a given Farmlet processes a fixed amount of data at each discrete time step. Three distinct types of errors are introduced randomly within each Worker DSP at a variable rate using a Multiply With Carry (RWC8gen) random number generator with a fixed seed. Any time a software or hardware error is encountered within the simulation, the processing rate for that DSP decreases a set amount depending on the type of error. The error is cleared when any DSP within the same Farmlet performs FVLA checks against the DSP for the error type present. However, there is a time cost associated with performing these checks. As detailed in the section above, the DSP must decide whether or not it is worth taking time to perform FVLA monitoring tasks against neighboring DSPs. If checks are performed too frequently, then the time available for data crossing processing is limited. On the other hand, if they are not performed frequently enough, then the chance that other DSPs within the same Farmlet are experiencing errors is high. As described, a high error rate will also lead to slow processing rates.

4.4.2 Utility Value Drives Real-Time Scheduling

The utility value formed in equation 4.1 is used to determine when to perform FVLA monitoring tasks, where

c_{ij} = 1 unit of data processing (20 microseconds), the fixed cost of DSP i performing FVLA monitoring tasks on DSP j,

b_{ij} = 10 units of data processing (200 microseconds), the fixed estimated benefit received by DSP i finding an error on DSP j, and

$p(t)$ = the adjusted sigmoid value for the number of time steps elapsed (t) since DSP i last performed the monitoring task on DSP j.

The utility value is calculated at each time step to determine whether the VLA or PA should have control of the DSP.

4.4.3 Self-* Emergent Behavior

Currently, a relatively rudimentary method is used to adapt the d-value (sensitivity) of each agent to a particular type of error. The approach relies only on a very limited amount of local information gathered when FVLA monitoring tasks are performed. As described in detail earlier, the d-value is modified when errors are found (not found) at each time step. In these experiments, the d-value is increased .005 when an error is found, and decreased .00001 if no error is found.

The DSPs self-organize as different DSPs within the Farmlet take on the necessary monitoring tasks at different points in time as required by the environment. In this way, the monitoring tasks required by the environment are always met, but not necessarily by one (or a few) designated DSPs. Instead, these tasks are performed by any polymorphic DSP within the Farmlet as dictated by the changing needs of the environment.

The results obtained show polymorphic VLAs evolving responsibility for a core set of fault monitoring tasks. Over the 100000 time steps for which the SWARM simulation is run, the 5 VLAs (1 per DSP) can be seen taking on distinct roles that lead to an efficient global fault mitigation strategy for monitoring errors on DSP1.

The simulation fluctuates the error rate at various intervals in order to demonstrate the affect changes in error rate can have on polymorphic behavior. A moderate error rate (5×10^{-4}) is used for the first 35000 time steps, a low error rate (5×10^{-6}) for the next 35000 time steps (35001-70000), and the last 30000 time steps (70001-100000) use a high rate (5×10^{-3}). The VLAs are able to adjust sensitivity to errors on DSP1 based on these fluctuating error rates over time. For example, the d-value (sensitivity) to individual errors on DSP1 for all 5 VLAs (embedded within DSP2 - DSP6) can be seen dropping beginning around time step 35000, and then increasing dramatically again at time step 70000 in reaction to the significant increase in error rate.

The VLA d-value (sensitivity) for 3 distinct error types was monitored on DSP1 within a single 6 DSP Farmlet. The d-values evolved by each of the VLAs within the 5 DSPs (DSP2-DSP6) monitoring DSP1 within the same Farmlet were evaluated. When the error rate is high (from time steps 70000-100000), the VLAs embedded within DSP3 and DSP6 develop a high sensitivity for error type 1 (e1), while the sensitivity for e1 of the VLAs in the remaining DSPs remains low. Similarly, the VLAs on DSP2 and DSP5 have a high sensitivity for error type 2 (e2), and VLAs for DSP2 and DSP3 are highly sensitive to e3.

The moderate error rate used for the first 35000 time steps reveals additional polymorphic behavior. Here, the error rate is not quite high enough for any single VLA to evolve long term responsibility for an individual error type on DSP1. Instead, 1 or 2 VLAs can be seen monitoring a single error type at one moment, and then a separate VLA (or group of VLAs) can be seen monitoring the same error type a short time later. This is due to the fact that the error rate is too low to stimulate high sensitivity in a single VLA. Sensitivity for the error type drops to a level comparable with other available VLAs on the Farmlet. For example, the VLAs on DSP 3 and DSP 4 develop a modest level of sensitivity for e1 early on (time steps 0-15000), but the role is then taken over by VLAs on DSP 5 and DSP 6 (time steps 15000-28000), and finally (time steps 28000-35000) taken back by VLAs on DSP 3 with a little help from VLAs on DSP2 and DSP 4.

4.5 Lessons Learned

There were a number of lessons learned while using the SWARM software development kit to simulate the RTES/BTeV environment. Many of these have to do with the way in which SWARM represents the concept of time. Simulations are run in discrete time steps, with every agent acting on all corresponding rules at each time step.

One of the first issues that we encountered is that the agents are by default evaluated in the exact same order at each time step. The problem that this causes is that since the set of errors are introduced at the initiation of each time step, the agents that are evaluated first each time will always be the first to encounter errors. Essentially this means that the same set of agents are always doing all of the work, while others sit idle. A DefaultOrder option was then found that could be set to 'randomized' so that the order of agent evaluation is random at each time step. Another challenge due to the constraint of time steps is how to accurately simulate a real-time environment, particulary one as large-scale as BTeV. A lot of work went into deciding how to accurately simulate processing and error rates as described above.

It was also found that simulation performance varries greatly between the software platforms that SWARM is designed to run on. Although the available GUI packages are far more robust with the Java version of SWARM, the Objective C version was used for these experiments due to the far better performance received during runtime. This was critical given the large number of agents, time steps, and hardware configurations that were simulated for the results obtained. SWARM also offers a batch mode which records system states directly to output files at each time step.

4.6 Next Steps

The next phase of this project will expand the SWARM simulation by increasing the number of different types of errors handled, along with the amount of fluctuation in error rates. Another important area of investigation for the next phase is to focus further on how sensitivity (d-value) to individual errors is adapted by each VLA. As described, a rudimentary method is currently used that slightly increases (or decreases) sensitivity based on the presence (or absense) of an error. Other variables could be considered in determining the amount of change to apply, such as factoring in the severity level of the error, or looking at the consequences of other recently taken actions. An enhanced evaluation methodology to better demonstrate the performance advantage of this approach as compared to other traditional methodologies is also necessary.

4.7 Summary

This chapter has described the details of a SWARM simulation of a fully distributed stigmergic approach to fault mitigation in large-scale real-time systems using light-weight, polymorphic, self-* agents embedded within 2500 individual DSPs. Stigmergy facilitates indirect communication and coordination between agents using cues from the environment, and concepts from game theory and social insect polymorphism allow individual agents to evolve a core set of roles for which it is responsible. Agents adapt these roles as environmental demands change. The approach is implemented on a SWARM simulation of RTES/BTeV, a data acquisition system for a High Energy Physics experiment consisting of 2500 DSPs.

The research conducted was sponsored by the National Science Foundation in conjunction with Fermi National Laboratories, under the BTeV Project, and in association with RTES, the Real-time, Embedded Systems Group. This work has been performed under NSF grant # ACI-0121658.

References

1. B. Barbat and C. Zamrescu. Polymorphic Agents for Modelling E-Business Users. International NAISO Congress on Information Science Innovations, Symposium on E-Business and Beyond (EBB), Dubai, 2000.
2. R. Burkhart. Schedules of Activity in the SWARM Simulation System. Position Paper for OOPSLA Workshop on OO Behavioral Semantics, 1997.
3. F. Cristian. Abstractions for fault-tolerance. In K. Duncan and K. Krueger, editors, Proceedings of the IFIP 13th World Computer Congress. Volume 3 : Linkage and Developing Countries, pages 278-286, Amsterdam, The Netherlands, 1994. Elsevier Science Publishers.
4. M. Daniels. An Open Framework for Agent-based Modeling. Applications of Multi-Agent Systems in Defense Analysis, a workshop held at Los Alamos Labs, April 2000.
5. J. Dowling, R. Cunningham, E. Curran, and V. Cahill. Component and system-wide self-* properties in decentralized distributed systems. Self-Star: International Workshop on Self-* Properties in Complex Information Systems, University of Bologna, Italy, May 31 - June 2 2004.
6. P. P. Grasse. La reconstruction du nid et les coordinations inter-individuelles chez Bellicosi-termes natalensis et Cubitermes sp. La theorie de la stigmergie: Essai d'interpretation des termites constructeurs. Insectes Sociaux, 6:pages 41-83, 1959.
7. W. Heimerdinger and C. Weinstock. A conceptual framework for system fault tolerance. Software engineering institute, carnegie mellon university, cmu/sei-92-tr-33, esc-tr-92-033, October, 1992.
8. N. M. Josuttis. Object Oriented Programming in C++. John Wiley & Sons; 1st edition, 2002.
9. J. Kowalkowski. Understanding and Coping with Hardware and Software Failures in a Very Large Trigger Farm. Conference for Computing in High Energy and Nuclear Physics (CHEP), March 2003.
10. S. Kwan. The BTeV Pixel Detector and Trigger System. FERMILAB-Conf-02/313, December 2002.
11. J. H. Law, W. O. Wilson, and J. McCloskey. Biochemical Polymorphism in Ants. Science, 149:pages 544-6, July 1965.
12. Z. Li, H. Liu, and M. Parashar. Enabling autonomic, self-managing grid applications.
13. D. Messie, M. Jung, J. Oh, S. Shetty, S. Nordstrom, and M. Haney. Prototype of Fault Adaptive Embedded Software for Large-Scale Real-Time Systems. 2nd Workshop on Engineering of Autonomic Systems (EASe), in the 12th Annual IEEE International Conference and Workshop on the Engineering of Computer Based Systems (ECBS), Washington, DC USA, April 2005.
14. J. Odell, H.V.D. Parunak, M. Fleischer, S. Breuckner. Modeling Agents and their Environment. Agent-Oriented Software Engineering III. Lecture Notes in Computer Science. volume 2585. Springer-Verlag. Berlin Heidelberg New York, 2002.
15. D. Messie and J. Oh. SWARM Simulation of Multi-Agent Fault Mitigation in Large-Scale, Real-Time Embedded Systems. High Performance Computing and Simulation (HPC&S) Conference, Magdeburg, Germany, June 2004.

16. J. Oh, D. Mosse, and S. Tamhankar. Design of Very Lightweight Agents for Reactive Embedded Systems. IEEE Conference on the Engineering of Computer Based Systems (ECBS), Huntsville, Alabama, April 2003.
17. D. Weyns, H. Parunak, F. Michel, T. Holvoet, and Jacques Ferber. Environments for Multiagent Systems, State-of-the-art and Research Challenges. Post-proceedings of the First International Workshop on Environments for Multiagent Systems, Lecture Notes in Artificial Intelligence, volume 3374, 2005.
18. D. E. Wheeler. Developmental and Physiological Determinants of Caste in Social Hymenoptera: Evolutionary Implications. American Naturalist, 128:pages13-34, 1986.
19. E. O. Wilson. The Origin and Evolution of Polymorphism in Ants. Quarterly Review of Biology, 28:pages 136-156, 1953.
20. M. Wooldridge and N. R. Jennings. Intelligent agents: Theory and Practice. Knowledge Engineering Review. volume 10 number 2 pages 115-152. citeseer.ist.psu.edu/article/wooldridge95intelligent.html. 1995.

Part II

Experiences Using Particle Swarm Intelligence

5

Swarm Intelligence — Searchers, Cleaners and Hunters * **

Yaniv Altshuler[1], Vladimir Yanovsky[1], Israel A. Wagner[1,2], and Alfred M. Bruckstein[1]

[1] Computer Science Department, Technion, Haifa 32000 Israel.
  {yanival, volodyan, wagner, freddy}@cs.technion.ac.il
[2] IBM Haifa Labs, MATAM, Haifa 31905 Israel.
  wagner@il.ibm.com

This chapter examines the concept of *swarm intelligence* through three examples of complex problems which are solved by a decentralized swarm of simple agents. The protocols employed by these agents are presented, as well as various analytic results for their performance and for the problems in general. The problems examined are the problem of finding patterns within physical graphs (e.g. *k-cliques*), the *dynamic cooperative cleaners* problem, and a problem concerning a swarm of UAVs (unmanned air vehicles), hunting an evading target (or targets). In addition, the work contains a discussion regarding open questions and ongoing and future research in this field.

5.1 Introduction

Significant research effort has been invested during the last few years in design and simulation of intelligent swarm systems. Such systems can generally be defined as decentralized systems, comprising relatively simple agents which are equipped with limited communication, computations and sensing abilities, designed to accomplish a given task ([8, 9, 10, 11]).

However, much work is yet to be done for such systems to achieve sufficient efficiency. This is caused, among others, by the fact that the geometrical theory of such multi agent systems (which is used to tie geometric features of the environments to the systems' behaviors) is far from being satisfactory, as pointed out in [12] and many others. Furthermore, while strong analysis

---

* This research was partly supported by the Ministry of Science Infrastructural Grant No. 3-942
** This research was partly supported by the Devorah fund

Y. Altshuler et al.: *Swarm Intelligence — Searchers, Cleaners and Hunters*, Studies in Computational Intelligence (SCI) **26**, 93–132 (2006)
www.springerlink.com   © Springer-Verlag Berlin Heidelberg 2006

of a swarm protocol and its performance is crucial for the development of stronger protocols and for overcoming the *resource allocation problem* users of multi agent system often face, most of the works in the field present merely a superficial analysis of the protocols, often in the sole form of experimental results.

This work introduces three complex problems which are to be solved by such intelligent swarms, utilizing specially designed protocols. The protocols and their performance are analyzed, and experimental results concerning the actual performance of the protocols are presented.

A more elaborated overview of previous work which concerns swarm intelligence is presented in Section 5.1.1 while details regarding the motivation behind this research appears in Section 5.1.2. A key element in design and analysis of swarm based systems and of swarm algorithms is the simplicity of the agents (in means on sensing and computation capabilities, communication, etc.). A discussion concerning this issue appears in Section 5.1.3.

Several works considered multi agents robotics in static environments. Such works can be found in [1], [2], [7] and elsewhere. These works present, among other results, protocols that assume the only changes taking place in the environment to be caused through the activity of the agents. The first problem presented in this work is a problem in which the agents must work in dynamic environments — an environment in which changes may take place regardless of the agents' activity. This problem is a dynamic variant of the known *Cooperative Cleaners* problem (described and analyzed in [1]). This problem assumes a grid, part of which is 'dirty', where the 'dirty' part is a connected region of the grid. On this dirty grid region several agents move, each having the ability to 'clean' the place ('tile', 'pixel' or 'square') it is located in, while the goal of the agents is to clean all the dirty tiles (in the shortest time possible). The dynamic variant of the problem (described in Section 5.2.1) involves a deterministic evolution of the environment, simulating a spreading *contamination* (or spreading *fire*).

The Dynamic Cooperative Cleaners problem was first introduced in [3], which also included a cleaning protocol for the problem, as well as several analytic bounds for the time it takes agents which use this protocol to clean the entire grid. A summary of those results appears in Section 5.2.4, while Section 5.2.5 describes a method of using a prior knowledge concerning the initial shape in order to improve its cleaning time.

The second problem presented is the *Cooperative Hunters* problem. This problem examines a scenario in which one or more *smart targets* (i.e. a platoon of T-72 tanks, a squad of soldiers, etc') are moving in a predefined area, trying to avoid detection by a swarm of UAVs (unmanned air vehicles). The UAV swarm's goal is to find the target (or targets) in the shortest time possible, meaning, we must guarantee that there exists time t in which all the escaping targets are detected, and that this t is minimal. While the swarm comprises relatively simple UAVs, which lack prior knowledge of the initial positions of the targets, the targets possess full knowledge of the whereabout of the swarm's agents, and are capable of intelligent evasive behavior.

A basic protocol for this problem and its analysis appears in [15]. However, this protocol assumes that the area in which the targets can move is known to the UAVs. Furthermore, although trying to minimize the communication between the swarm's agents, the work in [15] still requires a relatively high amount of explicit cooperation between the agents, which can be obtained through the use of a relatively high amount of communication.

This work contains a protocol for the requested task, which assumes no previous knowledge considering the area to be searched, and uses only a limited communication between the agents. The problem, the protocol and an analysis of its performance are presented in Section 5.3.

The third problem presented in this work is the *Physical k-Clique* problem, where a swarm comprising n mobile agents travels along the vertices of a physical graph G, searching for a *clique* of size k. A physical graph refers to a graph whose edges require a certain amount of time (or resources) for traveling along. Furthermore, information regarding the graph is available to the mobile agents only locally (meaning that the agents gather information only regarding the vertices they travel through). Thus, the complexity of the problem is measured in the number of edges traveled along, and not in the computational resources used by the agents. The work presents a search protocol for the agents, as well as experimental results for its performance (Section 5.4).

Section 5.5 contains a discussion regarding the problems, and details about extended research, which is currently being performed by the authors.

5.1.1 Swarm Intelligence — Overview

The area of multi agents and multi robotics distributed systems has become increasingly popular during the last two decades. Many applications, mainly in the contexts of computer networks, distributed computing and robotics, are nowadays being designed using techniques and schemes which are based on concepts derived from multi agents, or *swarms*, research.

The basic paradigm behind multi agents based system is that many tasks can be more efficiently completed by using multiple simple autonomous agents (robot, software agents, etc.) instead of a single sophisticated one. Regardless of the improvement in performance, such systems are usually much more adaptive, scalable and robust than those based on a single, highly capable, agent.

A multi agent system, or a swarm, can generally be defined as a decentralized group of multiple autonomous agents, either homogenous or heterogenous, such that those agents are simple and possess limited capabilities. Section 5.1.3 discusses the various limitations expected from such agents, while a commonly used taxonomy for multi agent robotics can be found in [47].

The inherent complexity of distributed multi agent systems, which is derived from the multitude of free variables involved in the operation and the decision process of such systems, makes their analysis extremely difficult. This may explain the relatively small number of theoretical results in this field, and the fact that most works in this field are justified through experimental results, or by analysis of simple cases only. Further hardness is derived from the fact that distributed multi agent systems are complex systems by nature, with all the classical features of such systems. Thus, the field of multi agent systems becomes an exciting and largely unexplored field for research and development.

Furthermore, while many works have been done considering multi agents in static environments (such as [1, 2, 7]), only a few works examined multi agent systems that operate in environments that change not only through the activity of the agents. As a results, the field of multi agent systems in dynamic environments is an exceptionally fascinating aspect of multi agents research.

Many research efforts have examined distributed systems models inspired by biology (see [55] or for behavior based control model — [64, 59], flocking and dispersing models — [78, 67, 69] and predator-prey approach — [61, 73]), physics [56], and economics [48, 49, 50, 51, 52, 53, 54].

Capabilities that have been of particular emphasis include task planning [57], fault tolerance [82], swarm control [79], human design of mission plans [77], role assignment [88, 65, 80], multi-robot path planning [89, 76, 70, 93], traffic control [84], formation generation [58, 96, 97], formation keeping [60, 91], target tracking [83] and target search [75].

Another interesting field of research is that of the biology inspired *ACO* metaheuristic for solving problems in dynamic environments. In [112], pheromones are treated as a *desirability feature* and are placed by the agents on the graph's edges. The information stored in these pheromones is essentially an implicit measurement of the probability of the edges to belong to the optimal solution for the problem (*TSP*, in that case). Other applications are -

- Sequential ordering problem [113].
- Quadratic assignment problem [114].
- Vehicle routing problem [115].
- Scheduling problem [116].
- Graph colouring problem [117].
- Partitioning problems [118].
- Problems in telecommunications networks [119], [120].

5.1.2 Swarm Intelligence — Motivation

The experience gained due to the growing demand for robotics solutions to increasingly complex and varied challenges has dictated that a single robot is no longer the best solution for many application domains. Instead, teams of robots must coordinate intelligently for successful task execution.

[99] presents a detailed description of multi robots application domains, and demonstrates how multi robots systems are more effective than a single

robot in many of these domains. However, when designing such systems it should be notice that simply increasing the number of robots assigned to a task does not necessarily improve the system's performance — multiple robots must cooperate intelligently to achieve efficiency.

Following are the main inherent advantages of a multi agent robotics (note that much of them hold for other multi agent systems, such as a distributed anti-virus mechanism, for example) :

- The benefit of parallelism — in task-decomposable application domains, robot teams can accomplish a given task more quickly than a single robot by dividing the task into sub-tasks and executing them concurrently. In certain domains, a single robot may simple no be able to accomplish the task on its own (e.g. carrying a large and heavy object).

- Robustness — generally speaking, a team of robots provides a more robust solution by introducing redundancy, and by eliminating any single point of failure. While considering the alternative of using a single sophisticated robot, we should note that even the most complex and reliable robot may suffer an unexpected malfunction, which will prevent it from completing its task. When using a multi agent system, on the other hand, even if a large number of the agents stop working from some reason, the entire group will often still be able to complete its task, albeit slower. For example, for exploring a hazardous region (such as a minefield or the surface of Mars), the benefit of redundancy and robustness offered by a multi agent system is highly noticeable.

- Adaptivity and Locality — the unit of a multi agents based systems has the ability of dynamically reallocating sub-tasks between the group of agents, thus adapting to unexpected changes in the environment. Furthermore, since the system is decentralized, it can respond relatively quickly to such changes, due to the benefit of locality, meaning — the ability to perform changes in the operation of a sub group of agents without the need to notify or request approval from any centralized "leader". Note that as the system comprises more agents, this advantage becomes more and more noticeable.

- Scalability — as in the previous section, as a multi agent system becomes larger, its relative performance in comparison to a centralized system becomes better. The scalability of multi agent systems is derived from the low overhead (both in communication and computation) such system possess. As the tasks assigned nowadays to multi agents based systems become increasingly complex, so does the importance of the high scalability of the systems.

- Heterogeneousness — since a group of agents may be heterogenous, it can utilize "*specialists*" — agents whose physical properties enable them to perform efficiently certain well defined tasks.

- Flexibility — as multi agent systems possess a great deal of internal complexity, such systems are capable of presenting a wide variety of behavior patterns. As a result, many kinds of tasks can be carried out by the same multi agent system. For example, a group of simple mobile robots can form a rigid line in order to scan an area for evading target, traverse an area in order to implement a "peeling" mechanism (such as [1] or [3]), or patrol an area in order to minimize the longest time between two visits in a certain point. This flexibility allows designers of such systems to use generic components and thus design fast and cheap systems for a given problem, whereas using a single robot requires designers to produce a special robot for each task.

Following are several examples for typical applications for which multi robotics system may fit successfully :

- Covering — in order to explore an enemy terrain, or clean a pre-defined minefield. May also be used for virtual spaces, such as a distributed search engine in a computer network [1, 3].
- Patrolling — for example, guarding a museum against thieves [121, 122].
- Cooperative attack, which require the cooperative and synchronized efforts of a large number of autonomous robots [15].
- Construction of complex structures and self-assembling (for example, see work on reconfigurable robots in [63, 62, 66, 86, 94, 72, 90, 95]).
- Missions which by nature require an extremely large number of extremely small units. For example, nano-robots performing medical procedures inside a human body [123, 124].
- Mapping and localizing — one example of this work is given in [71], which takes advantage of multiple robots to improve positioning accuracy beyond what is possible with single robots. Another example for a heterogenous localization system appears in [98].
- Environment manipulation — like a distributed system of transporting heavy payloads (see [85, 87, 68, 74, 92]).

5.1.3 Swarm Intelligence — Simplicity

A key principal in the notion of swarms, or multi agent robotics is the simplicity of the agents. Since "simplicity" is a relative description by its nature, the meaning is that the agents should be "much simpler" than a "single sophisticated agent" which can be constructed.

For example, cheap unmanned air vehicles which can fly at the speed of 200 MPH and detect targets at radius of 2 miles, and are able to broadcast to a range of 1 mile are "much simpler" than an F-16 airplane which can fly at a top speed of 1500 MPH, equipped with a radar which can detect targets at a range of 50 miles and use satellite based communication system. As technology advances, the criteria for future "simple agents" may of course be changed.

As a result, the resources of such simple agents are assumed to be limited, with respect to the following aspects :

- Memory resources — basic agents should be assumed to contain only $O(1)$ memory resources. This usually impose many interesting limitation on the agents. For example, agents can remember the history of their operation to only a certain extent. Thus, protocols designed for agents with such limited memory resources are usually very simple and try to solve the given problem by defining some (necessarily local) basic patterns. The task is completed by repetition of this patterns by a large number of agents.

   In several cases, stronger agents may be assumed, whose memory size is proportional to the size on the problem (i.e. $O(n)$). Such models may be used for example for problems when assuming that the agents are UAVs.

- Sensing capabilities — defined according to the specific nature of the problem. For example, for agents moving along a 100×100 grid, the sensing radius of the agents may be assumed to be 3, but not 40.

- Computational resources — although agents are assumed to employ only limited computational resources, a formal definition of such resources is hard to define. In general, most of the polynomial algorithms may be used.

Another aspect of swarms' and swarm algorithms' simplicity is the use of communication. The issue of communication in multi agent systems has been extensively studied in recent years. Distinctions between implicit and explicit communication are usually made, in which implicit communication occurs as a side effect of other actions, or "through the world" (see, for example [81]), whereas explicit communication is a specific act designed solely to convey information to other robots on the team.

Explicit communication can be performed in several ways, such as a short range point to point communication, a global broadcast, or by using some sort of distributed shared memory. Such memory is often treated to as a *pheromone*, used to convey small amounts of information between the agents [100, 101, 102]. This approach is inspired from the coordination and communication methods used by many social insects — studies on ants (e.g. [103, 104]) show that the pheromone based search strategies used by ants in foraging for food in unknown terrains tend to be very efficient.

Generally, we aspire that the agents will have as little communication capabilities as possible. Although a certain amount of implicit communication can hardly be avoided (due to the simple fact that by changing the environment, the agents are constantly generating some kind of implicit information), explicit communication should be strongly limited or avoided altogether. However, in several cases it can be shown that by adding a limited amount of communication to the agents' capabilities, much stronger systems can be produced.

5.2 The Dynamic Cooperative Cleaners (DCC) Problem

5.2.1 The Problem's Definition

Let us assume that the time is discrete. Let G be a two dimensional grid, whose vertices have a binary property of 'contamination'. Let $cont_t(v)$ state the contamination state of the vertex v in time t, taking either the value "on" or "off". Let F_t be the dirty sub-graph of G at time t, i.e. $F_t = \{v \in G \mid cont_t(v) = on\}$. We assume that F_0 is a single connected component. Our algorithm will preserve this property along its evolution.

Let a group of k agents that can move across the grid G (moving from a vertex to one of its 4-Connected neighbors in one time step) be arbitrarily placed in time t_0 on the boundary of F_0 (we focus on the cleaning problem, and not on the discovery problem). 4-Connected vertices are a pair of vertices who share a common border while 8-Connected vertices are vertices which share either common border or a common corner.

Each agent is equipped with a sensor capable of telling the condition of the square it is currently located in, as well as the condition of the squares in the $8 - Neighbors$ group of this square. An agent is also aware of other agents which are located in its square, and all the agents agree on a common "north". Each square can contain any number of agents simultaneously.

When an agent moves to a vertex v, it has the possibility of causing $cont(v)$ to become off. The agents do not have any prior knowledge of the shape or size of the sub-graph F_0 except that it is a single connected component.

Every d time steps the contamination spreads. That is, if $t = nd$ for some positive integer n, then $(\forall v \in F_t ,\ \forall u \in 4-Neighbors(v)\ :\ cont_{t+1}(u) = on)$.

The agents' goal is to clean G by eliminating the contamination entirely, so that $(\exists t_{success}\ :\ F_{t_{success}} = \emptyset)$. In addition, it is desired that this $t_{success}$ will be minimal.

In this work we demand that there is no central control and that the system is fully 'de-centralized' (i.e. all agents are identical and no explicit communication is allowed).

5.2.2 Solving the Problem — Cleaning Protocol

For solving the Dynamic Cooperative Cleaners problem the **SWEEP** cleaning protocol was suggested [3]. This protocol can be described as follows. Generalizing an idea from computer graphics (which is presented in [14]), the connectivity of the contaminated region is preserved by preventing the agents from cleaning what is called critical points — points which disconnect the graph of contaminated grid points. This ensures that the agents stop only upon completing their mission. An important advantage of this approach, in addition to the simplicity of the agents, is fault-tolerance — even if almost all the agents cease to work before completion, the remaining ones will eventually complete the mission, if possible. The protocol appears in Figure 5.2.

In the spirit of [13] we consider simple robots with only a bounded amount of memory (i.e. a *finite-state-machine*). At each time step, each agent cleans its current location (assuming this is not a critical point), and moves to its *rightmost* neighbor (starting from the agent's previous location — a local movement rule, creating the effect of a clockwise traversal of the contaminated shape). As a result, the agents "peel" layers from the shape, while preserving its connectivity, until the shape is cleaned entirely.

Since we only allow agents to clean boundary points, we guarantee that no new "holes" are created. The simple-connectivity of F, if such exists, is thus kept. This however, does not hold for a certain family of initial shapes. A complete discussion regarding a criteria for F_0 which guarantees that new holes will not be created can be found in [4] and a comprehensive work discussion this issue is currently under preparation by the authors. In general, the preservation of the simple-connectivity is rather easily guaranteed for digitally convex shapes lacking areas which may turn into holes once F spreads. An example of a shape in which the preservation of the simple-connectivity of the shape cannot be guaranteed appears in Figure 5.1.

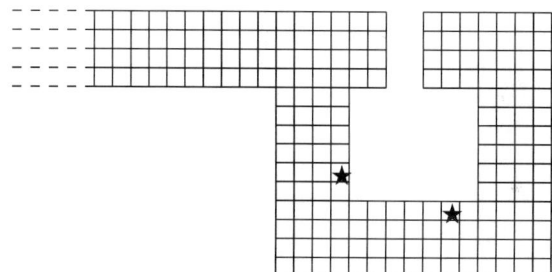

Fig. 5.1. A creation of a new hole around an agent. Notice that when the contamination spreads, it might trap one or more agents (denoted by stars) inside a "hole" that is created. In this case, the agents will continue cleaning the interior area of the hole.

5.2.3 Cleaning Protocol - Definitions and Requirements

Let $r(t) = (\tau_1(t), \tau_2(t), \ldots, \tau_k(t))$ denote the locations of the k agents at time t. The requested cleaning protocol is therefore a rule f such that for every agent i, $f(\tau_i(t), Neighborhood(\tau_i(t)), \mathcal{M}_i(t)) \in \mathcal{D}$, where for a square v, $Neighborhood(v)$ denotes the contamination states of v and its $8-Neighbors$, $\mathcal{M}_i$ is a finite amount of memory for agent i, containing information needed for the protocol (e.g. the last move) and $\mathcal{D} = \{'left', 'right', 'up', 'down'\}$.

Let ∂F_t denote the boundary of F_t. A square is on the boundary if and only if at least one of its $8-Neighbors$ is not in F_t, meaning $\partial F = \{(x, y) \mid (x, y) \in F \wedge 8-Neighbors(x, y) \cap (G \setminus F) \neq \emptyset\}$.

The requested rule f should meet the following goals :

- **Full Cleanness** : ($\exists t_{success}$: $F_{t_{success}} = \emptyset$). Notice that this demand is limited to cases where this is possible. Since we cannot know whether completing the mission is possible, we can only demand that when $d \to \infty$ the agents should achieve this goal.
- **Agreement on Completion** : within a finite time after completing the mission, all the agents must halt.
- **Efficiency** : in time and in agents' memory resources.

The requested rule should also be *fault tolerant*.

5.2.4 Dynamic Cooperative Cleaners — Results

Notice that analyzing the performance of the protocol is quite difficult. First, due to the preservation of the *critical points*, such points can be visited many times by the agents without being cleaned. Second, due to the dynamic nature of the problem the shape of the contaminated region can dramatically change during the cleaning process.

Another difficulty which arises is that of guaranteeing the completion of the cleaning by the agents. We must show that the agents are cleaning the contaminated region faster than the spreading of the last. Since given an instance of the problem, we know no easy way of knowing the minimal time it takes k agents to clean it, we cannot always foretell whether these k agents will be able to complete the mission successfully.

Let d denote the number of time steps between two contamination spreads. Then, for example, for any pair of values of d and k, let F_0 be a straight line of length $\lceil \frac{1}{8}d^2k^2 + dk + \frac{1}{2} \rceil$. Then, by applying the lower bound for the cleaning time (which was shown in [3]), we can see that the size of the shape is continually growing.

Thus, producing bounds for the proposed cleaning protocol is important for estimating its efficiency. Following is a summery of the bounds which were shown and proven in [3].

i. Given a contaminated shape F_0 with initial area of S_0, and k agents employing *any* cleaning protocol, following is a lower bound for S_t (the area of F in time t), and thus for the cleaning time of the agents :

$$S_{t+d} \geq S_t - d \cdot k + \sqrt{8 \cdot (S_t - d \cdot k) - 4}$$

where d is the number of time steps between spreads. Note that if $S_0 \gg d \cdot k$ then the sequence may be increasing and the agents cannot cease the fire. In addition, note that the bound is generic and applies regardless of the cleaning protocol used by the agents. The bound is based on the calculation of the maximal area which can be cleaned by k agents in d time steps, combined the minimal number of new contaminated squares which can be created ones F spreads. Additional details can be obtained in [3].

Protocol **SWEEP**(x, y) :

If (not **is-critical**(x, y)) and $((x, y) \in \partial F)$ and (there are no other agents in (x, y)) then

      Set $cont(x, y)$ to *off*;   */*Clean current position*/*

If (x, y) has no contaminated neighbors then **STOP**;

If (there are no other agents in (x, y)) or (the agent has the highest **priority** among agents in (x, y)) then

      If $\neg((x, y) \in \partial F)$ and in the previous time step the
      agent's square was in ∂F then

            /* Spread had occurred. Search for ∂F */
            Move in 90° counterclockwise to the previous
            movement and skip the rest of the protocol;

      If $\neg((x, y) \in \partial F)$ and in the previous time step the
      agent's square was not in ∂F then

            /* Keep seeking ∂F */
            Move on the same direction as in the previous
            movement and skip the rest of the protocol;

      If $(x, y) \in \partial F$ then

            Go to the *rightmost* neighbor of (x, y);

End **SWEEP**;

Function **is-critical**(x, y) :

If (x, y) has two contaminated squares in its $4-Neighbors$ which are not connected via a sequence of contaminated squares from its $8-Neighbors$ then

      Return **TRUE**

Else

      Return **FALSE**;

End **is-critical**;

Function **priority**(i) :

$(x_0, y_0) = \tau_i(t - 1)$;

$(x_1, y_1) = \tau_i(t)$;

Return $(2 \cdot (x_0 - x_1) + (y_0 - y_1))$;

End **priority**;

Procedure **INITIALIZE**() :

Choose a starting point on ∂F, p_0;

Put all the agents in p_0;

For $(i = 1; i \leq k; i + +)$ do

      Start agent i according to the **SWEEP** protocol;

      Wait 2 time steps;

End **INITIALIZE**;

Fig. 5.2. The **SWEEP** cleaning protocol. The protocol is used by agent i which is located in square(x, y). The term *rightmost* is defined as "starting from the previous boundary point scan the neighbors of (x, y) in a clockwise order until you find another boundary point"

ii. Following is an upper bound for $t_{SUCCESS}$, the cleaning time of k agents, using the **SWEEP** cleaning protocol, for a given contaminated shape F_0 :

$$t_{quick\_clean} \triangleq \frac{8(|\partial F_0| - 1) \cdot (W(F_0) + k)}{k} + 2k$$

If $t_{quick\_clean} \leq d$ then $t_{SUCCESS} = \lceil t_{quick\_clean} \rceil$. Otherwise $(t > d)$:
If F_0 is convex***, find the minimal t for which :

$$\sum_{i=d+1}^{t} \frac{1}{S_0 - 1 + 2\lfloor \frac{i}{d} \rfloor^2 + (c_0 + 2)\lfloor \frac{i}{d} \rfloor} \geq \gamma + \frac{8}{k} \cdot \lfloor \frac{t}{d} \rfloor$$

where

$$\gamma \triangleq \frac{8(k + W(F_0))}{k} - \frac{d - 2k}{|\partial F_0| - 1}$$

Otherwise $(t > d, F_0$ is not convex), find the minimal t for which :

$$\sum_{i=d+1}^{t} \frac{1}{S_0 - 1 + 2\lfloor \frac{i}{d} \rfloor^2 + (c_0 + 2)\lfloor \frac{i}{d} \rfloor} \geq$$
$$\geq \alpha + \frac{8}{2k}\sqrt{\beta + 4\left(\lfloor \frac{t}{d} \rfloor^2 + \lfloor \frac{t}{d} \rfloor\right)}$$

where

$$\alpha \triangleq 8 + \frac{8}{2k} - \frac{d - 2k}{|\partial F_0| - 1} \quad and \quad \beta \triangleq 2S_0 + 2c_0 - 1$$

In both cases $t_{SUCCESS} = t$.

In the above, d is the number of time steps between contamination spreads, c_0 is the circumference of F_0, S_0 is the area of F_0 and $W(F_0)$ denotes the maximal of the shortest distances between an internal square of F_0 and a non-critical square of ∂F_0.

This bound is produced by defining a "depth" of a shape as the maximal shortest path between an internal point of F and a non-critical point on the boundary of F. By showing that once an agent traverses F using the **SWEEP** protocol the depth of F is decreased, and by limiting the increase the the depth of F due to contamination spreads, an upper bound over the cleaning time is produced. Additional details can be obtained in [3].

iii. For a given contaminated shape F_0, an upper bound for the number of agents needed to apply the **SWEEP** protocol, for a successful cleaning of the shape, is derived from the cleaning time bound above.

A computer simulation, implementing the **SWEEP** cleaning protocol, was constructed. The simulation examined shapes of various geometric features,

---

*** Meaning that for every two squares in F_0 there is a $4-Neighbors$ "shortest path" between them, which is entirely in F_0.

and tracked the cleaning time of k agents ($k \in [1, 100]$) which used the **SWEEP** protocol. Following is a sample of the results, including the predictions of the lower and upper bounds. The sample includes a *"cross"* of size 2960, *"rooms"* of size 3959, and a random, semi-convex, shape of size 5000, where $d = 1000$:

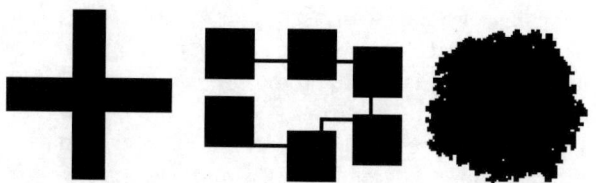

The lower curves represent the results predicted by the lower bound, while the upper curves represent the actual cleaning time, produced by the simulations performed (the graphs present the cleaning time as a function of the number of agents). The left graph presents the results that were produced by the "cross", etc'.

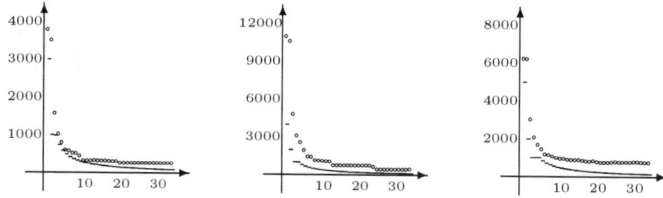

The following graph contains the upper bound for the "cross". Notice that the lower, tighter, curve was produced when taking into account that the "cross" shape is "convex":

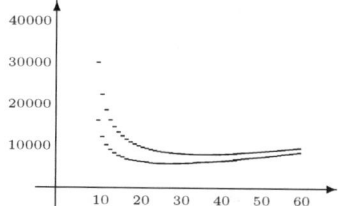

5.2.5 Dynamic Cooperative Cleaners — Convex Transformation

While [3] presented analytic upper bounds for the cleaning time of agents using the **SWEEP** protocol, assuming no prior knowledge concerning F_0 by the agents, in many cases such agents may indeed possess such information. Should this information be available to the agents, a method of employing it in order to improve the cleaning time should be devised.

While examining the simulation results, it can easily be seen that since the circumference of F_0 is rarely convex, valuable time is lost on cleaning the concave parts of it. Since our goal should be to minimize the size of the bounding sphere of the contaminated shape rather than the size of the contaminated shape itself, had the agents been aware of the existence of such parts, they shouldn't have bothered cleaning them. In order to satisfy this goal, when given information about F_0 the agents can calculate its convex hull (for example, using Graham scan) $C_0 \triangleq ConvexHull(F_0)$. Then, by using the **SWEEP** protocol over C_0 (instead of F_0) the agents are able to speed up the cleaning (since when C_0 is clean, so must be F_0). Using the upper bound of [3] allows us to predict this improvement.

Fig. 5.3 contains the results of the proposed transformation, for the *rooms* shape (size = 3959, circumference = 716) and the *random semi convex* shape (size = 5000, circumference = 654) of [3]. Since the *cross* which is presented in [3] is convex, the transformation will not change its cleaning time. The value of d for both shapes was chosen to be 1000. As can be seen, the use of convex transformation improved the performance of the agents by an average of approximately 35%.

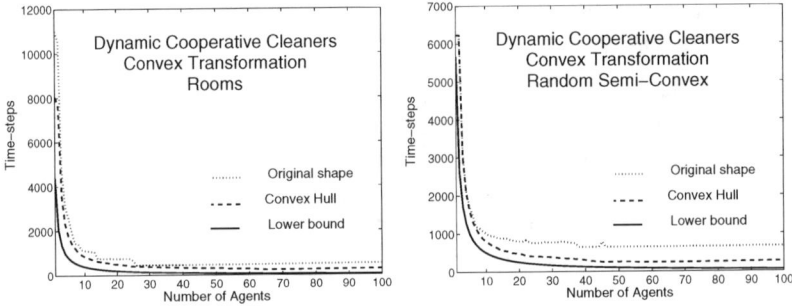

Fig. 5.3. Improvement in cleaning time of Convex-hull.

5.3 Cooperative Hunters

Significant research effort has been invested during the last few years in design, analysis and simulation of multi-agent systems design for searching areas (either known or unknown) [107, 108, 109, 110, 111]. While in most works the targets are assumed to be idle, recent works consider dynamic targets, meaning — target which by detecting the searching agents from a long distance, try to avoid detection by evading the agents.

Such problem is presented in [15], where a swarm of UAVs (unmanned aerial vehicles) is used for searching after one or more evading "smart targets". The UAVs swarm's goal is to guarantee that there exists time t in which all the escaping targets are detected, and that this t is minimal. In the current work, the problem described above will be denoted as the *Cooperative Hunters* problem.

5.3.1 Problem

Let us assume the time is discrete, while every time step lasts c_{time} seconds. Let G be a two dimensional grid, such that every vertex corresponds to an area in the map of size $c_{size} \times c_{size}$ square meters. Let us assume that each UAV moves at speed $1 \cdot c_{size}$ meters per time step. Let us assume that the targets move at speed $v_{target} \cdot c_{size}$ meters per time step (thus, c_{time} can be adjusted accordingly).

We assume that each UAV is equipped with sensors able of detecting the targets within its current vertex of G. The targets however, can spot the searcher from a great distance (considered to be infinite, and beyond the searcher's sensors range) and subsequently, manoeuver in order to evade detection. Once the searcher detects the target, it intercepts it.

Each UAV is aware of its current location (using a GPS receiver) and while flying over vertex v, can identify whether or not vertex v and its $8-Neighbors$ are part of the area to be searched. There is no limitation over the shape of the area to be searched, although it is assumed to be simply connected.

The UAV's communicate by using a wireless channel, while the information transmitted over this channel should be kept to a minimum.

The number of hiding targets can be either known to the swarms, or alternatively, the UAVs might not know the exact number of hiding targets (in which case the swarm will continue searching until guaranteeing that there is no longer a place in which the targets can hide). The goal of the swarm is to detect all hiding targets, in a minimal amount of time.

5.3.2 Motivation

The search strategy suggested in [15] for solving this problem defines *flying patterns* that the UAVs follow, which are designed for scanning the (rectangular) area in such a way that the targets cannot re-enter sub-areas which were already scanned by the swarm, without being detected by some UAV. Note that this protocol assumes that the area in which the targets can move is known to the UAVs in advance, and must be rectangular in shape.

However, this may not always be the case — a swarm of UAVs may be launched into an area whose shape and size is unknown to the UAVs prior to the beginning of the operation. For example, a swarm of UAVs might be used in order to hunt guerrilla forces, hiding in a city. These forces can be identified by the UAVs, since they use certain vehicles, or carrying certain electronic equipment or weapons, which can be identified by the UAVs' sensors. In this example, the targets can move only in the boundaries of the city, but the operators of the system may lack information regarding the exact boundaries of the city (since they may have only old and outdated satellite images of the area, or since large portions of the area was destroyed during recent fights). Thus, a method for searching without relying on previous knowledge of the searched area must be obtained.

The searching protocol presented in this work uses no prior knowledge of the area. The only assumption made is that the swarm's agents are capable of identifying the boundaries of the searched area, as they pass over them. In the previous example this means that once a UAV passes over the boundary of the city it can detect it. Therefore, a swarm whose agents contains no knowledge regarding the area to be searched, can still perform the searching task, by employing this protocol.

Furthermore, the presented protocol is very efficient in means of the simplicity of the agents, and the low communication between them. This makes the use of a simple protocol a notable advantage. While the protocol of [15] requires a relatively high amount of explicit cooperation between the UAVs (which can be obtained through the use of a relatively high amount of communication), the communication between the agents which is needed for a swarm using the presented protocol is extremely limited, and is bounded by $6 \cdot k$ bits per time step (k being the number of agents). Another advantage of the presented protocol is its fault tolerance, meaning — even if many UAVs malfunction or be shot down, the swarm will still be able to complete the task, albeit slower.

5.3.3 General Principle

Although the initial problem is that of searching for hiding targets within a given area, we shall consider an alternative, yet equivalent problem — the *dynamic cooperative cleaners* problem, presented in section 5.2.1.

Notice that from a cleaning protocol which is used by agents in order to solve the DCC problem, a protocol for the cooperative hunters problem can be derived. This is done by defining the entire area G as 'contaminated'. A 'clean' square (either a square which has not been contaminated yet, or a square which was cleaned by the UAVs) represents an area which is guaranteed not to contain any targets. By using the fact that the contamination is spreading, we simulate the fact that the targets may manoeuver around the UAVs, in order to avoid detection — if vertex v is contaminated then it may contain a target, thus, after $\frac{1}{v_{target}}$ seconds, this target could have moved from v to one of its neighbors, had it been in v. As result, after $\frac{1}{v_{target}}$ seconds all the neighbors of v become contaminated. In other words, the spreading contaminated simulates a *danger diffusion* that represents the capability of a square to contain a target.

The agents' goal is to eliminate the contaminated area — eliminate the places which the targets may be hiding in. Once there are no longer squares in which the targets may be hiding, the swarm is guaranteed to have detected all evading targets. Note that our demands regarding no prior knowledge of the search area are met, since the cooperative cleaners problem do not assume such knowledge.

5.3.4 Search Protocol

Let each UAV i hold G_i — a bitmap of G. Let every G_i be initialized to zeros (e.g. "clean"). Let each UAV i contain a hash table of vertices — f_i which for every vertex can return *on* or *off*. The default for all the vertices is *off*. The list f_i represents the vertices which are known to be within the area to be searched.

Every time a UAV flying over vertex v identifies v or one of its neighbors to be a part of the area to be searched, if $f_i(v) = off$ it sets the corresponding vertices of G_i to 1, sets $f_i(v)$ to be *on*, and broadcasts this information to the other UAVs. Once a UAV receives a transmission that vertex v is part of the area to be searched, it sets $f_i(v)$ to *on* and sets the corresponding vertex in G_i to 1. Every time a UAV moves it broadcasts the direction of its movement to the rest of the UAVs (north, south, west or east).

Notice that every time step each UAV broadcasts the new squares which are parts of G (which are set to 1 in G_i), and the squares it "cleaned" by flying over them (which are set to 0). Thus, the G_i and f_i of all UAVs are kept synchronized. Since v_{target} is known to the UAVs, they can simulate the spreading contamination, by performing ($\forall v \in G_i$, $\forall u \in Neighbors(v)$: $state(u) = 1$) every $\frac{1}{v_{target}}$ time steps. Thus, the bitmaps G_i always represent the correct representation of the area still to be cleaned.

The direction of movement and the decision whether or not to clean a vertex are determined using some cleaning protocol (for example, the **SWEEP** protocol of [3]). Notice that all the analytic bounds over the cleaning time of a cleaning protocol are immediately applicable for our hunting protocol. Whenever a UAV cleans a certain vertex, it sets this vertex in G_i to be 0, and broadcasts this information. Once a UAV receives such a transmission, it sets the vertex corresponding to the new location of the transmitting UAV to 0.

The UAVs are assume to be placed on the boundary of the area to be searched. Thus, each G_i immediately contains at least one vertex whose value is 1. As a result, for G_i to contains only zeros, the UAVs must have visited all the vertices of G and had made sure that no target could have escaped and "re-contaminated" a clean square. When G_i becomes all zeros UAV i knows that the targets have been found, and stops searching.

Since each time step, each UAV can move in at most 4 directions (i.e. 2 bits of information), clean at most a single vertex (i.e. 1 bit of information), and broadcast the status of 8 neighbor vertices (i.e. 3 bits of information), the communication is limited to 6 bits of information per UAV per time step.

5.3.5 Results

As mentioned in previous sections, by showing that any "cleaning protocol" may be used as the core component of a "hunting protocol" for the problem, all the analytic results concerning this cleaning protocol are immediately applicable. Specifically, by using the SWEEP cleaning protocol, all of the analytic bounds for its cleaning time (i.e. those mentioned in section 5.2.4 or in [3]) be utilized.

5.4 Physical k-Clique

5.4.1 Physical Graphs

The term *physical graph* denotes a graph $G(V, E)$ in which information regarding its vertices and edges is extracted using *I/O heads*, or *mobile agents*, instead of the "random access I/O" usually assumed in graph theory. These agents can physically move between the vertices of V along the edges of E, according to a predefined, or an on-line algorithm or protocol.

Moving along an edge e however, requires a certain amount of *travel efforts* (which may represent time, fuel, etc'). Thus, the complexity of algorithms which work on physical graphs is measured by the total travel efforts required, which equals the number of edges traveled by the agents. We assume that each edge requires exactly one unit of travel effort.

Physical graphs are conveniently used in order to represent many "real world problems", in which the most efficient algorithm is not necessarily the one whose computational complexity is the minimal, but rather one in which the agents travel along the minimal number edges. For example, the *Virtual Path Layout* problem, concerning the finding of a virtual graph of a given diameter, and its embedding in the physical graph such that the maximum load is minimized (see [36] and [37]).

Problems in physical graphs are thus variants of "regular" graph problems, such as finding a *k-clique* in a graph (description and algorithms can be found in [29]), graph and subgraph isomorphism (description and algorithms can be found in [17, 18, 19, 20, 21]), exploration problems (solved for example by *Breadth First Search* (BFS) and *Depth First Search* (DFS) algorithms [34]), etc., whose input graph is a physical graph. Thus, the complexity of these problems is measured as the number of edges an agent (or agents) solving the problem will travel along.

There is a special type of graph problems which can also be ideally described as physical graph problems. Such problems are those in which a *probabilistic*, or *real time* algorithm is used to return a solution which is not necessarily optimal. While a probabilistic algorithm returns a solution which is correct in a probability of $(1 - \epsilon)$ (for as small ϵ as we require), a real time algorithm can be asked at any stage to return a solution, whereas the quality of the solutions provided by the algorithm improves as time passes. Using such probabilistic or real time algorithms, the computational complexity of many problems can often be reduced from exponential to polynomial (albeit we are not guaranteed of finding the optimal solution). Such algorithms can be found for example in [22, 23, 24] (graph isomorphism) and [25, 31, 26] (distributed search algorithms such as *RTA**, *PBA**, *WS_PBA**, *SSC_PBA** and *BSA**). The physical variants of such problems can be thought of as a swarm of mobile agents, traveling the graph and collecting new information during this process. As time advances, more information is gathered by the agents, causing the quality of the solutions provided by the agents to improve.

Notice that while an algorithm which assumes a random access I/O model (from now on be referred to as *random access algorithm*) may read and write to the vertices of G at any order, an algorithm which assumes a physical data extraction (referred to as a *physical algorithm*) must take into account the distance between two sequential operations. The reason for this is that the use of a random access algorithm is performed using a processing unit and random access memory, whereas the use of a physical algorithm is actually done in the physical environment (or a simulated physical environment, which maintains the information access paradigm). Thus, a random access algorithm can access any vertex of the graph in $O(1)$, while a physical algorithm is confined to the distances imposed by the physical metric.

For example, for $u, v \in V$, let us assume that the distance between v and u in G is 5. Then if after a 'read' request from u, the algorithm orders a 'write' request to v, this process will take at least 5 time steps, and will therefore consume at least 5 fuel units. Furthermore, depending on the model assumed for the mobile agents knowledge base, this operation may take even longer, if, for example, the agents are not familiar with the shortest path from u to v, but rather know of a much longer path connecting the two vertices.

As can easily be seen from the previous example, while designing physical swarm algorithms, one must take into account an entire set of considerations which are often disregarded, while designing random access swarm algorithms.

5.4.2 The Problem — Pattern Matching by a Swarm of Mobile Agents

Pattern matching in graphs is the following problem: Given a graph G on n vertices and a graph H on h vertices, find whether G has an induced subgraph isomorphic to H. Many applications of pattern matching can be found in theory and practice of computer science, see e.g. [38] for more details. It is well known that this problem is computationally hard whenever H (and also h) is not constant, but rather a part of the input (a subgraph isomorphism problem), while it has a trivial solution of polynomial complexity if H is a constant graph. Note that the subgraph isomorphism problem is reducible to the *Max-Clique* problem.

This work considers dense physical graphs (graphs which contain many subgraphs isomorphic to the desired pattern) while the goal is to find one of them within as minimal moves on the graph as possible. A graph G on n vertices is called *dense* if the average degree of a vertex in G is $\Omega(n^2)$. Alternatively, we can slightly weaken this condition by requiring that the number of edges in G is $\Omega(n)$, as long as the existence of a large number of the requested patterns can be ensured. The vertices of G are indexed from 1 to n, and the edges of G are indexed from 1 to m, where $m \leq \binom{n}{2}$.

We assume that whenever a vertex $v \in V(G)$ is visited, all edges incident with v are revealed (i.e. their indices are revealed), and naturally v's index is also revealed. Hence, if an edge $e = (u, v)$ exists in G, and some agent visited u and v (in any order), then it can deduce that v is a neighbor of u, even if the agent did not travel on e. If there is a communication between the agents, it is enough that one of the agents visited u and some other agent visited v for this information to be deduced.

One of the artificial examples of a similar model might be an unknown terrain with indexed cities and roads, where the roads are signed with their indices (say with road signs), but their end points' indices are not mentioned. However, we do not assume that the graph is planar (i.e. there might be roads, bridges and tunnels, crossing each other in any way).

Similar to ordinary navigation tasks (for example [30, 31, 32, 33]), the purpose of each agent employing the search protocol is to reach the "goal vertex" as soon as possible. However, since the goal of the agents is to detect a *k-clique*, the goal vertex is in fact the vertex which when discovered completes a *k-clique* in the agent's knowledge base. Thus, there is no specific goal vertex, but rather several *floating goal vertices*, whose identities depends on the following elements :

- The structure of G (namely, V and E).
- The information the agent had collected thus far.
- The information sharing model of the problem (be it a distributed memory, centralized memory, etc').

Note also that this problem is not an ordinary exploration problem (see [16]), where the entire graph should be explored in order for it to be mapped out. Once a requested *k-clique* is found by one of the agents, the problem is terminated successfully. This often occurs while only a small portion of the graph's vertices have been visited.

Another distinction should be made between the problem that is presented in this work and those examined in the field of multi agents routing (see [39, 40]). While multi agents routing mainly deals with the problem of finding paths in dynamic networks where the structure of the environment is dynamically changing due to load balancing and congestion, the *physical k-clique* problem considers a stable physical environment (somewhat similar to the work of [7] and [2]).

5.4.3 Motivation

It is our belief that work on swarm protocols for physical graphs is strongly required since physical graphs and networks, which have an essential role in nowadays research and industry application, are becoming more and more complex, and thus new techniques for such graphs must be composed. Several examples for such networks are the *world wide web* [35, 28, 27], the physical and logical infrastructure of the internet [28], power grids, electronic circuits [28] — all of which are complex physical environments.

The *physical k-clique* problem has several "real world" applications. For example, tracking the connectivity of a computer or telephone network, which can be utilized in order to improve routing schemes within this network. Another example may be a distributed mechanism which searches many databases containing transactions and email correspondences, in order to identify a group of people who maintain tight connections between the group's members (a possible indicator of a potential terrorists group).

Another reason the *physical k-clique* problem was selected for this research is that the *k-clique* problem, on which the *physical k-clique* problem is based, is known to be a significantly hard problem. While most of known NP-complete problems can be approximated quite well in polynomial (and sometimes even linear) time (as shown in [125]), this is not the case for the *k-clique* problem. An explanation of why there are no "semi-efficient" algorithms for the *k-clique* problem (and thus, that the best solution for it is an exhaustive search) and why the *k-clique* problem can not be approximated in *any* way, unless $P = NP$, can be found in [126]. Additional details regarding NP-Complete problems can be found in [127].

Since the *physical k-Clique* problem is in fact an instance of the *physical pattern matching* problem, solving it can serve as a first step towards a general pattern matching swarm algorithm. In future works we intend to show that the algorithm presented in this work can be applied to any pattern with few modifications.

5.4.4 Physical Clique Finding Protocol

Let $r(t) = (\tau_1(t), \tau_2(t), \ldots, \tau_n(t))$ denote the locations of the n agents at time t. The requested search protocol is therefore a movement rule f such that for every agent i, $f(\tau_i(t), N(\tau_i(t)), \mathcal{M}_i(t)) \in N(\tau_i(t))$, where for a vertex v, $N(v)$ denotes the neighbors of v, (e.g. $N(v) \triangleq \{u \in V \mid (v, u) \in E\}$).

$\mathcal{M}_i$ is the memory for agent i, containing information gathered by it through movement along G and by reading information from the shared memory. The requested rule f should meet the following goals :

* **Finding a k-Clique** : $(\exists t_{success} \mid k\text{–}clique \in \mathcal{M}_i(t_{success}))$ such that this $t_{success}$ is minimal.

- **Agreement on Completion** : within a finite time after the completion of the mission, all the mobile agents must be aware that the mission was completed, and come to a halt.
- **Efficiency** : in time, agents' memory and the size of the shared memory.

The requested rule should also be *fault tolerant*, meaning that even if some of the agents malfunction, the remaining ones will eventually find the target clique, albeit somewhat slower.

For solving the *Physical k-Clique* problem a search protocol named **PCF** is suggested. This protocol can be described as follows: each agent holds a knowledge base which contains some information regarding G. Every time an agent enters a vertex v, it performs a *data merge* process, in which the knowledge base of the agent updates and is updated by the central knowledge base.

The main idea of the search protocol is exploring the graph in directions that have the highest potential for discovering the desired pattern. Considering only cliques simplifies the arguments because of the clique's perfect symmetry.

All potential sets of vertices are sorted according to the largest clique which is contained in the set. Sets containing the same size of maximal clique are sorted according to the total number of unexplored edges which touch the vertices of the set (unexplored edges are edges $e(v, u)$ whereas at least one the identities of v and u is yet unknown) . As large the number of such edges is, the more likely it is for the set to be expandable to a *k-clique*. In addition, if a *k-clique* was found, the sort criteria places it on the top of the list, hence the agents are immediately aware of it. The algorithm uses a *job list*, containing the sets described above.

Generally, when looking for a pattern H of size h in graph G, every set of $m, m < h$ visited vertices in G that might be completed to a subgraph of G isomorphic to H (i.e. there are enough unexplored edges for every vertex) forms a potential sub-H of size m. While considering cliques as the patterns, we only need to verify that the m vertices of the set form a clique and that every vertex in the set has at least $h - (m + 1)$ unexplored edges (which is the minimal requirement in order for this set to be expandable to an $h - clique$. Sets which do not meet this demand are deleted from the job list. The job list is updated (if needed) after every move of the agents.

If there are α available agents in a turn, the algorithm assigns the top α jobs in the sorted job list to the agents, where the assignments are made in a way that minimizes the total travel distance in L_∞ norm. This is done in order to minimize the travel efforts of the agents. This is an example of a difference between physical problems and "regular" problems — whereas in conventional complexity scheme all the ways of dispersing the jobs among the agents are identical, in the complexity scheme of physical problems we would like to assign jobs to the nearest agents.

Once an agent reaches its goal (the closest vertex of the job assigned to the agent), it writes the new discovered information in the adjacency matrix and becomes available. At the beginning of each turn, if the first job in the list is associated to a list of k vertices the algorithm declares that a clique has been found, and terminates. Notice that all the agents use a common job list. In addition, the agents are assumed to broadcast every new information they encounter, creating a *common shared memory*.

We assume that all distances (i.e. graph edges) are equal to one unit, and that the agents have sufficient memory and computational power. In addition, we assume that that there are many cliques of the required size in G (otherwise, if the number of cliques is too small (e.g. $o\binom{n}{k}$), the optimal algorithm for discovering them would be the exhaustive exploration of G). The algorithm, used by each agent i appears in Figure 5.4.

Algorithm **PCF** :
While the first job in the common job list is not associated to a set of k vertices
       Assign top job in jobs list to the agent;
       Perform the job;
       Update the agent's data structures;
       Update the agent's jobs list (or the common list);
       Update the list of available agents;
       If (all vertices explored) then
           STOP;
End **PCF**;

Procedure **CREATE DATA STRUCTURES**(i) :
Create an $n \times n$ matrix with entries of type integer;
// *Represents edge indices according to vertex adjacency of G,*
// *as discovered by the agents.*
Create a sorted list of jobs;
// *Each job is associated with one potential clique.*
// *The job list is sorted according to the following criteria :*
// *(a) Size of the potential sub clique*
// *(b) Number of still unexplored edges*
// *(c) The distance of the job from the agents*
// *The distance is computed according to L_∞ norm, that is the*
// *min—max on the travel distance.*
End **CREATE DATA STRUCTURES**;

Procedure **INITIALIZE**() :
Initialize the agents' jobs lists;
// *Note that the common job list may contain large potential*
// *sub-cliques, immediately upon initialization.*
CREATE DATA STRUCTURES(i);
Choose random starting points on G for the agents;
Place the agents in these starting points;
Start the agents according to the **PCF** algorithm;
End **INITIALIZE**;

Fig. 5.4. The **PCF** search algorithm for the centralized shared memory model.

5.4.5 Results

The algorithm was tested on *Erdös-Renyi* random graphs $G \sim G(n, p)$ where G has n vertices, and each pair of vertices forms an edge in G with probability p independently of each other. In order to ensure that G has enough clique sub-graphs of size k, the value of p should be chosen wisely. Formally, the probability that k vertices of G form a clique is p^k, thus by linearity of expectation and the second moment, G will have at least $\frac{1}{4}\binom{n}{k}p^k$ cliques of size k with probability $1 - o(1)$ (for further background on probabilistic arguments see e.g. [46]). Since we require that G will contain a sufficient number of k-clique (namely, $O(\sqrt{n})$ k-cliques), we choose the value of p with respect to the formula above, and specifically :

$$p = \left(\frac{\sqrt{16n} \cdot k!}{\prod_{i=0}^{k-1} n - i} \right)^{\frac{1}{k}}$$

Fig. 5.5 contains experimental results of the search algorithm for cliques of size 10 in graphs of sizes 500 and 2000. The number of agents searching the graphs is 5 through 50 agents. It can be seen that while adding more agents dramatically improves the search time, the system reach a state of saturation, after which adding more agents yields only mild contribution to the performance of the swarm, if any. This can be best observed in the smaller graph, where enlarging the number of agents from 15 to 50 decreases the search time by less than 30%, where as increasing the number of agents from 5 to 15 decreases the search time by 70%.

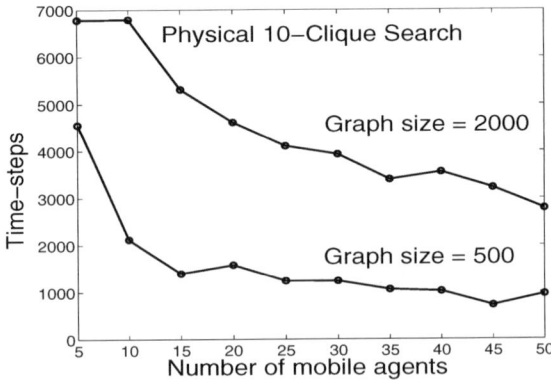

Fig. 5.5. Results of the Physical 10-Clique problem. Notice how the performance of the swarm increases while adding more agents, until it reaches a state of saturation.

Fig. 5.7 examines the increase in the swarm's search time for larger graphs. Notice that unlike the regular *k-clique* problem, in which the search time

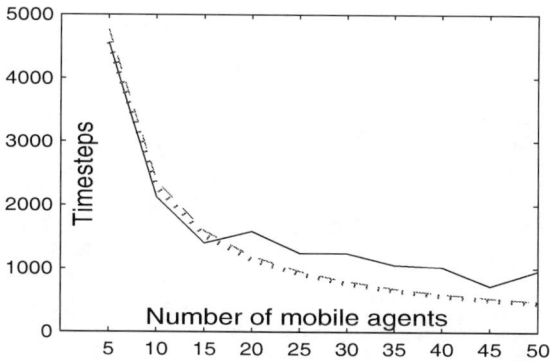

Fig. 5.6. A comparison between the results of the Physical 10-Clique problem in a graph of 500 vertices as a function of the number of agents, and the behavior of the $y = \frac{1}{x}$ function. The dotted curve presents the $y = \frac{1}{x}$ function where $y(5)$ equals the experimental solving time of 5 agents. The red dashed curve presents the instance of $y = \frac{1}{x}$ function which obtains the minimal distance from the entire set of experimental results, in L_2 norm. The proximity of the curves indicates that the system operates as expected. Observe how the later points of the experimental results are located above the $y = \frac{1}{x}$ curve, which can be expected since the utilization of the system is likely to be smaller than 100%.

equals $O(n^k)$, the results are almost linear in n. This demonstrates the basic claim that once computation resources are disregarded in comparison of travel effort, the complexity of many central problems in computer science change dramatically.

In addition, note that the search time for 20 agents is consistently smaller than for 10 agents. However, while in smaller graphs the difference is approximately 50%, for larger graphs it shrinks down to 25% for a graph of 4500 vertices. Interestingly, when increasing the number of agents to 30, the performance almost does not improve at all (meaning that the system had reached a state of saturation).

5.4.6 Exploration in Physical Environments

While considering the physical k-clique problem, an interesting discussion can take place considering the comparison between this problem and the problem of physical exploration by a swarm of mobile agents. The exploration problem for a decentralized group of mobile agents is merely the problem of achieving a complete knowledge of the graph under investigation by one of the agents. This comparison is indeed interesting since under the assumptions of a physical environment (namely, that computation resources can be disregarded in comparison to travel efforts) once an exploration of the environment

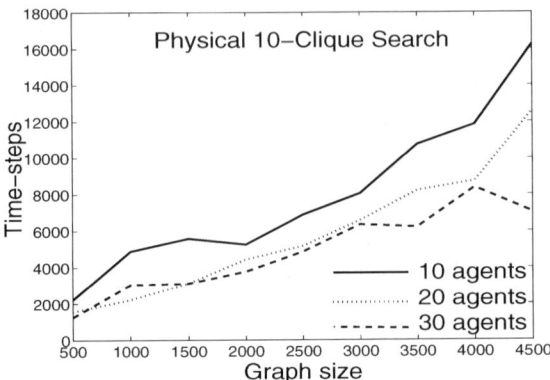

Fig. 5.7. Results of the Physical 10-Clique problem. Notice how the performance of the swarm increases while increasing the number of agents from 20 to 30, whereas this number is increased to 30, the performance almost never changed. In addition, observe the linearity of the graph, whereas the complexity of the problem in orthodox graph theory scheme is $O(n^k)$.

is completed, all cliques contained in the world are revealed. As a result, upper bounds for the exploration problem serve also as upper bounds for the physical k-clique problem.

Let G be a physical graph over n vertices. We assume that our physical environments are *Erdös-Renyi* random graphs $G \sim G(n,p)$ where G has n vertices, and each pair of vertices form an edge in G with probability p independently of each other. The edge density parameter of G, p may be either a constant or a function of n. We also assume that the agents are spread uniformly at random over the nodes of G. We shall assume without loss of generality, that the agents move according to the random walk algorithm. This type of behavior simulates the weakest model of agents, and hence time bounds here may impose similar time bounds for other essential behaviors of agents (since an agent can always act according to the random walk algorithm, and achieve its exploration time).

Under the above settings, we are interested in examining the number of required steps on the physical graph G, after which the whole graph G is known to some agent (or similarly, the complete information regarding the graph G is stored in one of its nodes). Let m denote the number of agents.

In order to obtain an upper bound for the exploration time of a random walkers swarm in G let us use the following bound of [129] :

$$E(ex_G) = O\left(\frac{|E|^2 log^3(|V|)}{m^2}\right)$$

Results of this bound for graphs used by the k-clique search protocol appears in Figures 5.8 and 5.9.

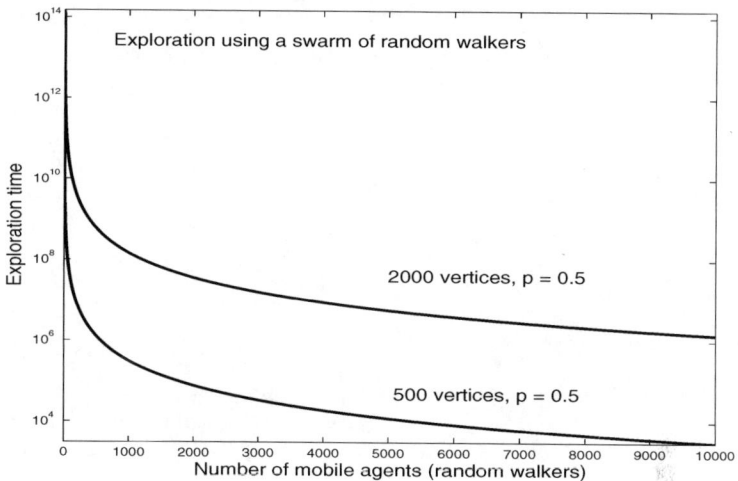

Fig. 5.8. Results predicted by the bound of [129] regarding exploration of general graphs using a swarm of mobile agents which use the random walk algorithm. The graph shows the mission completion time as a function of the number of agents.

5.4.7 Swarm Intelligence for Physical Environments — Related Work

Hitherto, there have been various works which examine problems in physical graphs, as well as the use of swarm based systems for such problems. For example, [40, 41] use mobile agents in order to find shortest paths in (dynamically changing) telecom networks, where the load balancing on the edges of the graph is unstable. Similarly, several widely used Internet routing algorithms (such as *BGP* [42], *RIP* [43] etc') propagate shortest path information around the network and cache it at local vertices by storing routing tables with information about the next hop. Another known routing system which uses "ants" and "pheromonoes" is the *AntNet* system [39]. In *AntNet*, ants randomly move around the network and modify the routing tables to be as accurate as possible.

While these approaches seem similar, there is a great difference between these works and one presented here, both concerning the environment in which the agent operate, the data that is stored at each vertex and the problem to be solved.

First, most of the works mentioned above which concern routing problems, assume that the environment is a telecom network of some sort, which changes dynamically over time. In this work, we assume the graph to be fixed and corresponds to a real physical environment (with Euclidean distances or some other metric), while the difficulty is derived from fact that the graph is not known to the agents.

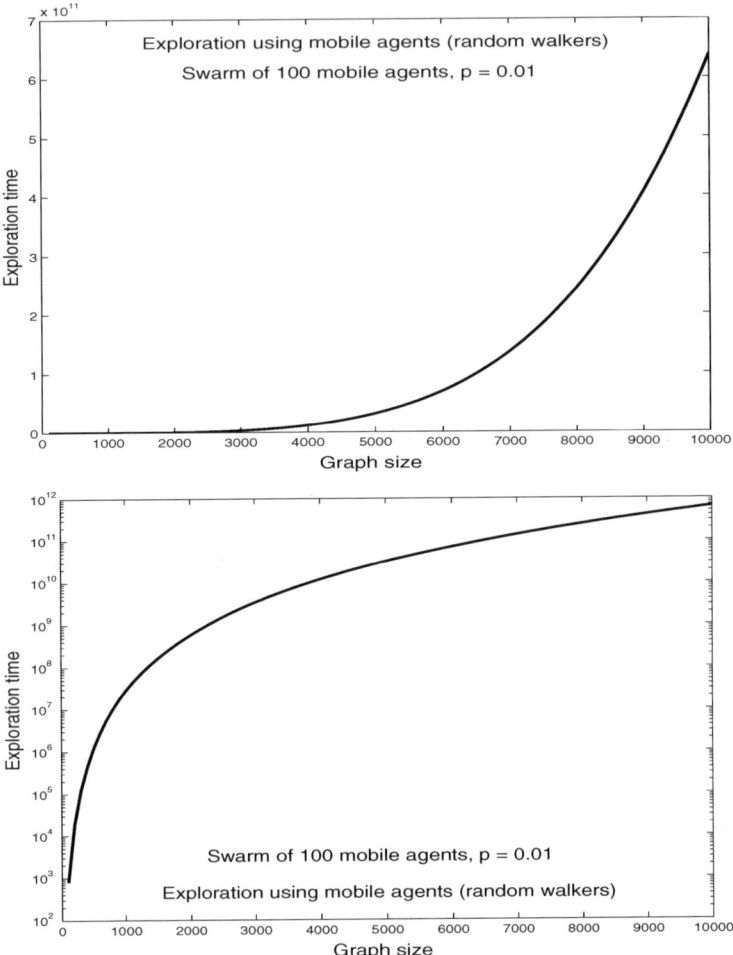

Fig. 5.9. Results predicted by the bound of [129] regarding exploration of general graphs using a swarm of mobile agents which uses the random walk algorithm. The graphs show the mission completion time as a function of the size of the graph.

Another difference is that these algorithms try to help a mobile agent (which for example, represents a phone conversation) to navigate to a certain goal. In other words, given the destination vertex (which could be any vertex of the environment) and the knowledge written in the current vertex, these algorithms try to locally decide where should the agent go to next, while the goal of each agent is merely to minimize the time it takes this particular agent to reach its destination. In the *physical k-clique* problem, on the other hand, the agents' goal is to find k vertices which form a *k-clique*, while the goal of the agent is to minimize the time it takes the entire *swarm* to find such a clique.

Essentially, the approach presented in this work can be seen as a generalization of the *next hop lookup tables* mentioned earlier, since according to the partial knowledge of the graph, an agent decides on its next move, when this process is continually advancing, until a clique is found.

Similar to this approach, the works of [44] and [45] present a mechanism which find the shortest path within a physical graph, while assuming several communication mechanism.

5.5 Discussion and Conclusion

In this work three problems in the field of swarm intelligence were presented. These problems have several "real world" applications. While the Cooperative Hunters problem is already formulated in the form of such a problem, the Cooperative Cleaners problem may be used, for example, for a coordinating fire-fighting units, or an implementation of a distributed anti-virus system for computer networks. Additional applications are distributed search engines, and various military applications (such as a basis for UAV swarm systems, as offered in Section 5.3.3). Regarding the physical k-Clique problem, many of its applications were discussed in Section 5.4.3. Protocols for the problems were presented and analyzed, and several results were shown.

In addition, one of the major principles considering works in physical environments was shown (in Figure 5.6 of Section 5.4.5), in which a problem whose time complexity (according to orthodox complexity principles) equals $O(n^k)$ presented a physical complexity of only $O(n)$, demonstrating a basic difference between the two complexity schemes.

While examining these problems, new interesting opportunities for an extended research have emerged. We have already started investigating some of the above, producing more valuable results. Following are various aspects of this ongoing and future research :

5.5.1 Cooperative Cleaners

One of the important features of the SWEEP cleaning protocol is its simplicity. As mentioned in great length in Section 5.1.3, our paradigm assumes the use of highly simple agents, and thus derives designs of highly simple protocols. While examining the performance of this protocol, one may wonder what is the price which we pay by adopting such simple protocols instead of other more complex and resource demanding protocols. In order to answer this question, the authors have been experiencing with an $A*$ based mechanism (see [106]) which (after much centralized and highly computation demanding exhaustive processes) is able of producing *optimal solutions* for cleaning problems (meaning, solutions which are guaranteed to produce cleaning within the shortest time possible). Surprisingly, the results of these experiments have shown that the performance of the **SWEEP** protocol are only roughly 30%–50% slower than the optimal solutions. These amazing results successfully demonstrate that a fully decentralized simple swarm protocol, assuming no communication or global knowledge, can produce results which are extremely close to the optimal solutions of the problem under investigation. This is possible thanks to the swarm behavior which emerges through the use of such protocol. The results of these experiments were published in [6] and an example appears in Figures 5.10 and 5.11.

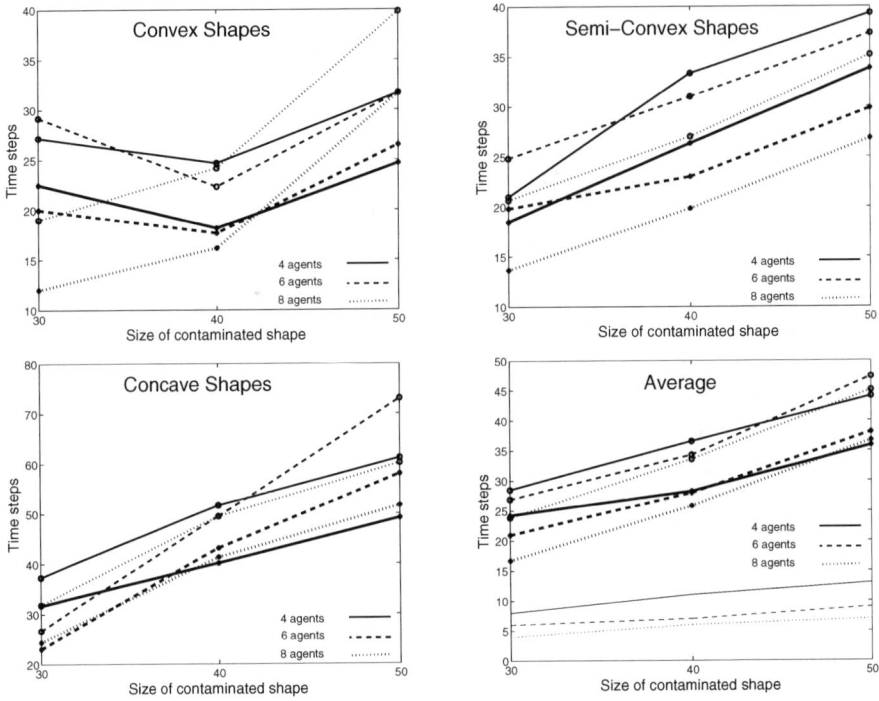

Fig. 5.10. Comparison of sub optimal and optimal algorithms. The lower and ticker three lines represent the cleaning time of the optimal algorithm whereas the upper lines represents the **SWEEP** cleaning protocol. In the right chart on the bottom, the lower three lines represent the lower bound of the optimal solution, as appears in Section 5.2.4.

Another interesting aspect is the feasibility question, i.e. foretelling the minimal number of agents required to clean a given shape (regardless of the cleaning time). In addition, developing new cleaning protocols for the problem and producing tighter bounds are also of interest to us. The authors

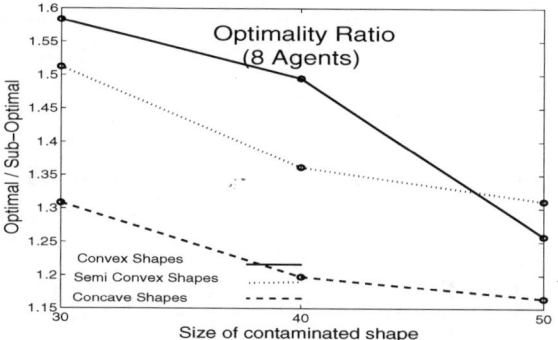

Fig. 5.11. Comparison of sub optimal and optimal algorithms. The Y-axes represents the $\frac{performance_{optimal}}{performance_{sub-optimal}}$ ratio for various sizes. Note that sa the problem is getting bigger, the sub-optimal performance of the **SWEEP** cleaning protocol get closer to those of the optimal algorithm.

are currently in the stages of developing improved cleaning protocols for the problem.

An explicit upper bound for the cleaning time of agents using the **SWEEP** protocol has been derived by the authors, using the Ψ function (see for example [105]). Another important result is the discovery of new geometric features, which are invariants with respect to the agents' activity and the spreading contamination. Their relevance to the problem was demonstrated by developing an improved upper bound for the cleaning time of the agents. Both results, including several other analytic results for the problem are presented in a paper which is currently under preparation.

Since many of the interesting "real world" applications take place in environments not easily reduced to lattices (for example, a distributed anti-virus mechanism for computer networks), additional analytic work should be performed concerning the dynamic cooperative cleaners problem in *non-planar graphs*. Although some of the existing work can immediately be applicable for such environments (for example, the generic lower bound, mentioned in Section 5.2.4), while trying to reconstruct other elements, one may face interesting new challenges.

5.5.2 Cooperative Hunters

While a flexible and fault-tolerant protocol for solving the problem, based on the dynamic cleaning problem was presented, there is still room for protocols, which will present enhanced performances. The authors are currently constructing hunting protocols which are not "cleaning protocols based" in hope of demonstrating improved results. One such protocol, which was shown to produce near-optimal results was recently constructed, and can be seen in [5].

In addition, there arises the question of robustness, meaning — given a swarm comprised of UAVs which are slightly slower than the minimal speed required by various hunting protocols, what is the average decrease in hunting performance of such swarm ? Thus, the quality of a hunting protocol is derived not only by the minimal speed limit it enforces on the UAVs or by the number of UAVs required, but also by its sensitivity to deviation from these values. A paper discussing the above is currently under preparation by the authors.

5.5.3 Physical k-Clique

While this work had assumed a *centralized shared memory* for the agents' communication, a second variant of the protocol, assuming a *distributed shared memory* model, in which the mobile agents can store and extract information using the graph's vertices is described by the authors in in[128]. This version is also much more similar to other common models of "real world" problems, since assuming storage of small portion of data in the graph nodes is considered much more easy assumption than assuming that the agents has the ability of broadcasting throughout the entire swarm. Surprisingly, our experimental results show that the performance of the special swarm search protocol designed for this problem is very close to the performance of the decentralized protocol. Again, this is a successful demonstration that the power of swarm intelligence may often compensate for shortage of sensing or communication capabilities.

In addition, since the protocol was designed to be employed by a swarm of mobile agents, it is highly dependent on the information gathered by the agents. In order to increase robustness, a special kind of *exploratory agents* were introduced to the system. A paper discussing the distributed swarm model as well as various enhancements (such as exploratory agents) and a many experimental results is currently under preparation.

As mentioned earlier, although the protocol described in this work was designed for the physical k-clique problem, we believe that by minor adjustments a generic protocol for finding a given pattern in a physical graph may be composed. The main difference between cliques and graph patterns in general is that cliques (or alternatively independent sets) have a perfect symmetry between their vertices. In other words, every pair of vertices in a clique has equivalent elements under some automorphism. The notion of "potential" is also naturally extensible to the general case. For example, given a pattern H on h vertices and a subset C of c vertices from the graph G, we say that C is potentially expandable to H if :

   i. $c < h$.

   ii. There exists an assignment π to the unknown pairs of vertices.

   iii. There exists a set C' of $(h - c)$ vertices such that the induced subgraph on $C \cup C'$ under the assignment π is isomorphic to H.

To scale the potential of a certain subset C, a good estimation of the amount of possible assignments under which C is expandable to a set of vertices that induce a subgraph isomorphic to H must be produced.

In addition, it is our intention to devise such a protocol, as well as a protocol for the pattern matching problem in *dynamic* physical graphs, a problem which also bears a great deal of interest.

References

1. I.A. Wagner, A.M. Bruckstein: "Cooperative Cleaners: A Case of Distributed Ant-Robotics", in "Communications, Computation, Control, and Signal Processing: A Tribute to Thomas Kailath", Kluwer Academic Publishers, The Netherlands (1997), pp. 289–308
2. I.A. Wagner, M. Lindenbaum, A.M. Bruckstein: "Efficiently Searching a Graph by a Smell-Oriented Vertex Process", Annals of Mathematics and Artificial Intelligence, Issue 24 (1998), pp. 211–223
3. Y. Altshuler, A.M. Bruckstein, I.A. Wagner: "Swarm Robotics for a Dynamic Cleaning Problem", in "IEEE Swarm Intelligence Symposium" (SIS05), Pasadena, USA, (2005)
4. Y. Altshuler, V. Yanovski: "Dynamic Cooperative Cleaners — Various Remarks", Technical report, CS-2005-12, Technion - Israel Institute of Technology, (2005).
5. Y.Altshuler, V.Yanovsky, I.A.Wagner, A.M. Bruckstein: "The Cooperative Hunters – Efficient Cooperative Search For Smart Targets Using UAV Swarms", Second International Conference on Informatics in Control, Automation and Robotics (ICINCO), the First International Workshop on Multi-Agent Robotic Systems (MARS), Barcelona, Spain, (2005).
6. Y.Altshuler, I.A.Wagner, A.M. Bruckstein: "On Swarm Optimality In Dynamic And Symmetric Environments", Second International Conference on Informatics in Control, Automation and Robotics (ICINCO), the First International Workshop on Multi-Agent Robotic Systems (MARS), Barcelona, Spain, (2005).
7. R.C. Arkin, T. Balch: "Cooperative Multi Agent Robotic Systems", Artificial Intelligence and Mobile Robots, MIT/AAAI Press, Cambridge, MA, (1998)
8. R.A. Brooks: "Elephants Don't Play Chess", Designing Autonomous Agents, P. Maes (ed.), pp. 3–15, MIT press / Elsevier, (1990)
9. S.Levi: "Artificial Life - the Quest for a New Creation", Penguin, (1992)
10. S.Sen, M. Sekaran, J. Hale: "Learning to Coordinate Without Sharing Information", Proceedings of AAAI-94, pp. 426–431
11. L.Steels: "Cooperation Between Distributed Agents Through Self-Organization", Decentralized A.I. — Proc. of the first European Workshop on Modeling Autonomous Agents in Multi-Agents world, Y.DeMazeau, J.P.Muller (Eds.), pp. 175–196, Elsevier, (1990)
12. G.Beni, J.Wang: "Theoretical Problems for the Realization of Distributed Robotic Systems", Proc. of 1991 IEEE Internal Conference on Robotics and Automation, pp. 1914–1920, Sacramento, California April (1991)
13. V.Breitenberg: Vehicles, MIT Press (1984)
14. D.Henrich: "Space Efficient Region Filling in Raster Graphics", The Visual Computer, pp. 10:205–215, Springer-Verlag, (1994)

15. P.Vincent, I.Rubin: "A Framework and Analysis for Cooperative Search Using UAV Swarms", ACM Simposium on applied computing, 2004
16. M.A. Bender, A. Fernandez, D. Ron, A. Sahai, S.P. Vadhan: "The power of a pebble: Exploring and mapping directed graphs", In the proceedings of the Thirtieth Annual ACM Symposium on the Theory of Computating, pp. 269–278, Dallas, Texas, May (1998).
17. L.P. Cordella, P. Foggia, C. Sansone, M. Vento: "Evaluating Performance of the VF Graph Matching Algorithm", Proc. of the 10th International Conference on Image Analysis and Processing, IEEE Computer Society Press, pp. 1172–1177, (1999).
18. J.R. Ullmann: "An Algorithm for Subgraph Isomorphism", Journal of the Association for Computing Machinery, vol. 23, pp. 31–42, (1976).
19. M.R. Garey, D.S. Johnson: "Computers and Intractability: A Guide to the Theory of NPCompleteness", Freeman & co., New York, (1979).
20. B.D. McKay: "Practical Graph Isomorphism", Congressus Numerantium, 30, pp. 45–87, (1981).
21. D.G. Corneil, C.C. Gotlieb: "An efficient algorithm for graph isomorphism", Journal of the Association for Computing Machinery, 17, pp. 51–64, (1970).
22. W.J. Christmas, J. Kittler, M. Petrou: "Structural Matching in Computer Vision Using Probabilistic Relaxation", IEEE Trans. Pattern Analysis and Machine Intelligence, vol. 17, no. 8, pp. 749–764, (1995).
23. P. Kuner, B. Ueberreiter: "Pattern recognition by graph matching — combinatorial versus continuous optimization", Int. J. Pattern Recognition and Artif. Intell., vol. 2, no. 3, pp. 527–542, (1988).
24. K.A. De Jong, W.M. Spears: "Using genetic algorithms to solve NP-complete problems", in Genetic Algorithms, (J.D. Schaffer, ed.), Morgan Kaufmann, Los Altus, CA., pp. 124–132, (1989).
25. A.A. Toptsis, P.C. Nelson: "Unidirectional and Bidirectional Search Algorithms", IEEE Software, 9(2), (1992).
26. J.B.H. Kwa: "BS*: An Admissible Bidirectional Staged Heuristic Search Algorithm", Artificial Intelligence, pp. 95–109, Mar., (1989).
27. D.J. Watts: "Small Worlds", Princeton University Press, Princeton NJ, (1999).
28. S.N. Dorogovtsev, J.F.F. Mendes: "Evolution of networks", Adv. Phys. 51, 1079, (2002).
29. D.S. Hochbaum, O. Goldschmidt, C. Hurken, G. Yu: "Approximation algorithms for the k-Clique Covering Problem", SIAM J. of Discrete Math, Vol 9:3, pp. 492–509, August, (1996).
30. P. Cucka, N.S. Netanyahu, A. Rosenfeld: "Learning in navigation: Goal finding in graphs", International journal of pattern recognition and artificial intelligence, 10(5):429–446, (1996).
31. R.E. Korf: "Real time heuristic search", Artificial intelligence, 42(3):189–211, (1990).
32. L. Shmoulian, E. Rimon: "Roadmap A*: an algoritm for minimizing travel effort in sensor based mobile robot navigation", In the proceedings of the IEEE International Conference on Robotics and Automation, pp. 356–362, Leuven, Belgium, May (1998).
33. A. Stentz: "Optimal and efficient path planning for partially known environments.", In the proceedings of the IEEE International Conference on Robotics and Automation, pp. 3310–3317, San Diego, CA, May (1994).

34. R.J. Wilson: "Introduction to Graph Theory", Longman, London, 2nd ed., (1979).
35. R. Albert, A.L. .Barabasi,: "Statistical Mechanics of Complex Networks", Reviews of Modern Physics, vol. 74, January, (2002).
36. O. Gerstel, S. Zaks: "The Virtual Path Layout problem in fast networks", In Proceedings of the Thirteenth Annual ACM Symposium on Principles of Distributed Computing, pp. 235–243, Los Angeles, California, August, (1994).
37. S. Zaks: "Path Layout in ATM networks", Lecture Notes in Computer Science, 1338:pp. 144–177, (1997).
38. A. Apostolico, Z. Galil: "Pattern Matching Algorithms", Oxford University Press, Oxford, UK.
39. G. Di Caro, M. Dorigo: "AntNet:Distributed stigmergetic control for communiction networks", Journal of Artificial Intelligence Research, 9:317–365, (1998).
40. S. Appleby, S. Steward: "Mobile software agents for control in telecommunication networks", British Telecom Technology Journal, 12, pp. 104–113, (1994).
41. R. Schnooderwoerd, O. Holland, J. Bruten, L. Rothkrantz: "Ant-based load balancing in telecommunication networks", Adaptive Behavior 5(2), (1996).
42. Y. Rekhter, T. Li: "A Border Gateway Protocol", Request for Comments 1771, T.J Watson Research Center IBM Corporation & cisco Systems, March (1995).
43. G. Malkin: "RIPng Protocol Applicability Statement", RFC 2081, IETF Network Working Group, January, (1997).
44. A. Felner, R. Stern, A. Ben-Yair, S. Kraus, N. Netanyahu: "PHA*: Finding the Shortest Path with A* in Unknown Physical Environments", Journal of Artificial Intelligence Research, vol. 21, pp. 631–679, (2004)
45. A. Felner, Y. Shoshani, I.A.Wagner, A.M. Bruckstein: "Large Pheromones: A Case Study with Multi-agent Physical A*", Forth International Workshop on Ant Colony Optimization and Swarm Intelligence, (2004)
46. N. Alon, J. H. Spencer: "The probabilistic method", Wiley-Interscience (John Wiley & Sons), New York, (1992) (1st edition) and (2000) (2nd edition).
47. G. Dudek, M. Jenkin, E. Milios, D. Wilkes: "A Taxonomy for Multiagent Robotics". Autonomous Robots, 3:375397, (1996).
48. B.P.Gerkey, M.J.Mataric: "Sold! Market Methods for Multi-Robot Control", IEEE Transactions on Robotics and Automation, Special Issue on Multi-robot Systems, (2002).
49. M.Golfarelli, D.Maio, S. Rizzi: "A Task-Swap Negotiation Protocol Based on the Contract Net Paradigm", Technical Report, 005-97, CSITE (Research Center For Informatics And Telecommunication Systems, associated with the University of Bologna, Italy), (1997).
50. G.Rabideau, T.Estlin, T.Chien, A.Barrett: "A Comparison of Coordinated Planning Methods for Cooperating Rovers", Proceedings of the American Institute of Aeronautics and Astronautics (AIAA) Space Technology Conference, (1999).
51. R.Smith: "The Contract Net Protocol: High-Level Communication and Control in a Distributed Problem Solver", IEEE Transactions on Computers C-29 (12), (1980).
52. S.M.Thayer, M.B.Dias, B.L.Digney, A.Stentz, B.Nabbe, M.Hebert: "Distributed Robotic Mapping of Extreme Environments", Proceedings of SPIE, Vol. 4195, Mobile Robots XV and Telemanipulator and Telepresence Technologies VII, (2000).

53. M.P.Wellman, P.R.Wurman: "Market-Aware Agents for a Multiagent World", Robotics and Autonomous Systems, Vol. 24, pp.115–125, (1998).

54. R.Zlot, A.Stentz, M.B.Dias, S.Thayer: "Multi-Robot Exploration Controlled By A Market Economy", Proceedings of the IEEE International Conference on Robotics and Automation, (2002).

55. R.C.Arkin, T.Balch: "AuRA: Principles and Practice in Review", Journal of Experimental and Theoretical Artificial Intelligence, Vol. 9, No. 2/3, pp.175–188, (1997).

56. D.Chevallier, S.Payandeh: "On Kinematic Geometry of Multi-Agent Manipulating System Based on the Contact Force Information", The 6th International Conference on Intelligent Autonomous Systems (IAS-6), pp.188–195, (2000).

57. R.Alami, S.Fleury, M.Herrb, F.Ingrand, F.Robert: "Multi-Robot Cooperation in the Martha Project", IEEE Robotics and Automation Magazine, (1997).

58. T.Arai, H.Ogata, T.Suzuki: "Collision Avoidance Among Multiple Robots Using Virtual Impedance", In Proceedings of the IEEE/RSJ International Conference on Intelligent Robots and Systems, pp. 479-485, (1989).

59. R.C.Arkin: "Integrating Behavioral, Perceptual, and World Knowledge in Reactive Navigation", Robotics and Autonomous Systems, 6:pp.105-122, (1990).

60. T.Balch, R.Arkin: "Behavior-Based Formation Control for Multi-Robot Teams", IEEE Transactions on Robotics and Automation, December (1998).

61. M.Benda, V.Jagannathan, R.Dodhiawalla: "On Optimal Cooperation of Knowledge Sources", Technical Report BCS-G2010-28, Boeing AI Center, August (1985).

62. G.Beni: "The Concept of Cellular Robot", In Proceedings of Third IEEE Symposium on Intelligent Control", pp.57-61, Arlington, Virginia, (1988).

63. H.Bojinov, A.Casal, T.Hogg: "Emergent Structures in Moduluar Self-Reconfigurable Robots", In Proceedings of the IEEE International Conference on Robotics and Automation, pp.1734-1741, (2000).

64. R.A.Brooks: "A Robust Layered Control System for a Mobile Robot", IEEE Journal of Robotics and Automation, RA-2(1):14-23, March (1986).

65. C.Candea, H.Hu, L.Iocchi, D.Nardi, M.Piaggio: "Coordinating in Multi-Agent RoboCup Teams", Robotics and Autonomous Systems, 36(2- 3):67-86, August (2001).

66. A.Castano, R.Chokkalingam, P.Will: "Autonomous and Self-Sufficient CONRO Modules for Reconfigurable Robots", In Proceedings of the Fifth International Symposium on Distributed Autonomous Robotic Systems (DARS 2000), pp. 155-164, (2000).

67. J.Deneubourg, S.Goss, G.Sandini, F.Ferrari, P.Dario: "Self-Organizing Collection and Transport of Objects in Unpredictable Environments", In Japan-U.S.A. Symposium on Flexible Automation, pp.1093-1098, Kyoto, Japan, (1990).

68. B.Donald, L.Gariepy, D.Rus: "Distributed Manipulation of Multiple Objects Using Ropes", In Proceedings of IEEE International Conference on Robotics and Automation, pp.450=457, (2000).

69. A.Drogoul J.Ferber: "From Tom Thumb to the Dockers: Some Experiments With Foraging Robots", In Proceedings of the Second International Conference on Simulation of Adaptive Behavior, pp.451-459, Honolulu, Hawaii, (1992).

70. C.Ferrari, E.Pagello, J.Ota, T.Arai: "Multirobot Motion Coordination in Space and Time", Robotics and Autonomous Systems, 25:219-229, (1998).

71. D.Fox, W.Burgard, H.Kruppa, S.Thrun: "Collaborative Multi-Robot Exploration", Autonomous Robots, 8(3):325-344, (2000).

72. T.Fukuda, S.Nakagawa: "A Dynamically Reconfigurable Robotic System (Concept of a System and Optimal Configurations)", In Proceedings of IECON, pp.588-595, (1987).

73. T.Haynes, S.Sen: "Evolving Behavioral Strategies in Predators and Prey", In Gerard Weiss and Sandip Sen, editors, Adaptation and Learning in Multi-Agent Systems, pp.113-126. Springer, (1986).

74. O.Khatib, K.Yokoi, K.Chang, D.Ruspini, R.Holmberg, A.Casal: "Vehicle/Arm Coordination and Mobile Manipulator Decentralized Cooperation", In IEEE/RSJ International Conference on Intelligent Robots and Systems, pp.546-553, (1996).

75. S.M.LaValle, D.Lin, L.J.Guibas, J.C.Latombe, R.Motwani: "Finding an Unpredictable Target in a Workspace with Obstacles", In Proceedings of the 1997 IEEE International Conference on Robotics and Automation (ICRA-97), pp.737-742, (1997).

76. V.J.Lumelsky, K.R.Harinarayan: "Decentralized Motion Planning for Multiple Mobile Robots: The Cocktail Party Model", Autonomous Robots, 4(1):121-136, (1997).

77. D.MacKenzie, R.Arkin, J.Cameron: "Multiagent Mission Specification and Execution", Autonomous Robots, 4(1):29-52, (1997).

78. M.J.Mataric: "Designing Emergent Behaviors: From Local Interactions to Collective Intelligence", In J.Meyer, H.Roitblat, and S.Wilson, editors, Proceedings of the Second International Conference on Simulation of Adaptive Behavior, pp.432-441, Honolulu, Hawaii, MIT Press, (1992).

79. M.J.Mataric: "Interaction and Intelligent Behavior", PhD Thesis, Massachusetts Institute of Technology, (1994).

80. E.Pagello, A.DAngelo, C.Ferrari, R.Polesel, R.Rosati, A.Speranzon: "Emergent Behaviors of a Robot Team Performing Cooperative Tasks", Advanced Robotics, 2002.

81. E.Pagello, A.DAngelo, F.Montesello, F.Garelli, C.Ferrari: "Cooperative Behaviors in Multi-Robot Systems Through Implicit Communication", Robotics and Autonomous Systems, 29(1):65-77, (1999).

82. L.E.Parker: "ALLIANCE: An Architecture for Fault-Tolerant Multi-Robot Cooperation", IEEE Transactions on Robotics and Automation, 14(2):220-240, (1998).

83. L.E.Parker, C.Touzet: "Multi-Robot Learning in a Cooperative Observation Task", In Distributed Autonomous Robotic Systems 4, pp.391-401. Springer, (2000).

84. S.Premvuti, S.Yuta: "Consideration on the Cooperation of Multiple Autonomous Mobile Robots", In Proceedings of the IEEE International Workshop of Intelligent Robots and Systems, pp.59-63, Tsuchiura, Japan, (1990).

85. D.Rus, B.Donald, J.Jennings: "Moving Furniture with Teams of Autonomous Robots", In Proceedings of IEEE/RSJ International Conference on Intelligent Robots and Systems, pp.235-242, (1995).

86. D.Rus M.Vona: "A Physical Implementation of the Self-Reconfiguring Crystalline Robot", In Proceedings of the IEEE International Conference on Robotics and Automation, pp.1726-1733, (2000).

87. D.Stilwell, J.Bay: "Toward the Development of a Material Transport System Using Swarms of Ant-Like Robots", In Proceedings of IEEE International Conference on Robotics and Automation, pp.766-771, Atlanta, GA, (1993).

88. P.Stone, M.Veloso: "Task Decomposition, Dynamic Role Assignment, and Low-Bandwidth Communication for Real-Time Strategic Teamwork", Artificial Intelligence, 110(2):241-273, June (1999).

89. P.Svestka, M.H.Overmars: "Coordinated Path Planning for Multiple Robots", Robotics and Autonomous Systems, 23(3):125-152, (1998).

90. C.Unsal, P.K.Khosla: "Mechatronic Design of a Modular self-Reconfiguring Robotic System", In Proceedings of the IEEE International Conference on Robotics and Automation, pp.1742-1747, (2000).

91. P.K.C.Wang: "Navigation Strategies for Multiple Autonomous Mobile Robots", In Proceedings of the IEEE/RSJ International Conference on Intelligent Robots and Systems (IROS), pp.486-493, (1989).

92. Z.Wang, Y.Kimura, T.Takahashi, E.Nakano: "A Control Method of a Multiple Non-Holonomic Robot System for Cooperative Object Transportation", In Proceedings of Fifth International Symposium on Distributed Autonomous Robotic Systems (DARS 2000), pp.447-456, (2000).

93. A.Yamashita, M.Fukuchi, J.Ota, T.Arai, H.Asama: "Motion Planning for Cooperative Transportation of a Large Object by Multiple Mobile Robots in a 3D Environment", In Proceedings of IEEE International Conference on Robotics and Automation, pp.3144-3151, (2000).

94. M.Yim, D.G.Duff, K.D.Roufas: "Polybot: a Modular Reconfigurable Robot", In Proceedings of the IEEE International Conference on Robotics and Automation, pp.514-520, (2000).

95. E.Yoshida, S.Murata, S.Kokaji, K.Tomita, H.Kurokawa: "Micro Self-Reconfigurable Robotic System Using Shape Memory Alloy", In Proceedings of the Fifth International Symposium on Distributed Autonomous Robotic Systems (DARS 2000), pp.145-154, (2000).

96. J.Fredslund, M.J.Mataric: " Robot Formations Using Only Local Sensing and Control", In the proceedings of the International Symposium on Computational Intelligence in Robotics and Automation (IEEE CIRA 2001), pp.308–313, Banff, Alberta, Canada, (2001).

97. N.Gordon, I.A.Wagner, A.M.Bruckstein: "Discrete Bee Dance Algorithms for Pattern Formation on a Grid", In the proceedings of IEEE International Conference on Intelligent Agent Technology (IAT03), pp.545–549, October, (2003).

98. R.Madhavan, K.Fregene, L.E.Parker: "Distributed Heterogenous Outdoor Multi-Robot Localization", In the proceedings of IEEE International Conference on Robotics and Automation (ICRA), pp.374–381, (2002).

99. M.B.Dias, A.Stentz: "A Market Approach to Multirobot Coordination": Technical Report, CMU-RI - TR-01-26, Robotics Institute, Carnegie Mellon University, (2001).

100. V. Yanovski, I.A. Wagner, A.M. Bruckstein: "A distributed ant algorithm for efficiently patrolling a network", Algorithmica, 37:165–186, (2003).

101. I.A. Wagner, A.M. Bruckstein: "ANTS: agents, networks, trees and subgraphs", Future Generation Computer Computer Systems Journal, 16(8):915–926, 2000.

102. V. Yanovski, I.A. Wagner, A.M. Bruckstein: "Vertex-ants-walk: a robust method for efficient exploration of faulty graphs. Annals of Mathematics and Artificial Intelligence, 31(1–4):99–112, (2001).

103. F.R. Adler, D.M. Gordon: "Information collection and spread by networks of partolling agents", The American Naturalist, 140(3):373–400, (1992).
104. D.M. Gordon: "The expandable network of ant exploration", Animal Behaviour, 50:372–378, (1995).
105. M. Abramowitz, I.A. Stegun: "Handbook of Mathematical Functions", National Bureau of Standards Applied Mathematics Series 55, (1964)
106. Hart, P.E., Nilsson, N.J., Raphael, B.: "A formal basis for the heuristic determination of minimum cost paths", IEEE Transactions on Systems Science and Cybernetics 4(2): 100–107, (1968).
107. Passino, K., Polycarpou, M., Jacques, D., Pachter, M., Liu, Y., Yang, Y., Flint, M. and Baum, M.: "Cooperative Control for Autonomous Air Vehicles", In Cooperative Control and Optimization, R. Murphey and P. Pardalos, editors. Kluwer Academic Publishers, Boston, (2002).
108. Polycarpou, M., Yang, Y. and Passino, K.: "A Cooperative Search Framework for Distributed Agents", In Proceedings of the 2001 IEEE International Symposium on Intelligent Control (Mexico City, Mexico, September 5–7). IEEE, New Jersey, 1–6, (2001).
109. Stone, L.D: "Theory of Optimal Search", Academic Press, New York, (1975).
110. Koopman, B.O: "The Theory of Search II, Target Detection", Operations Research 4, 5, 503–531, October, (1956).
111. Koenig, S., Liu, Y.: "Terrain Coverage with Ant Robots: A Simulation Study", AGENTS'01, May 28–June 1, Montreal, Quebec, Canada, (2001).
112. Dorigo M., L.M. Gambardella: "Ant Colony System: A Cooperative Learning Approach to the Traveling Salesman Problem", IEEE Transactions on Evolutionary Computation, 1(1):53-66 (1997).
113. Gambardella L. M. and M. Dorigo: "HAS-SOP: An Hybrid Ant System for the Sequential Ordering Problem", Tech. Rep. No. IDSIA 97-11, IDSIA, Lugano, Switzerland, (1997).
114. Gambardella L. M., E. Taillard and M. Dorigo: "Ant Colonies for the Quadratic Assignment Problem". Journal of the Operational Research Society, 50:167-176 (1999).
115. Bullnheimer B., R.F. Hartl and C. Strauss: "An Improved Ant system Algorithm for the Vehicle Routing Problem", Paper presented at the Sixth Viennese workshop on Optimal Control, Dynamic Games, Nonlinear Dynamics and Adaptive Systems, Vienna (Austria), May 21-23, (1997), appears in: Annals of Operations Research (Dawid, Feichtinger and Hartl (eds.): Nonlinear Economic Dynamics and Control, (1999)
116. Colorni A., M. Dorigo, V. Maniezzo and M. Trubian: "Ant system for Job-shop Scheduling", JORBEL - Belgian Journal of Operations Research, Statistics and Computer Science, 34(1):39-53 (1994).
117. Costa D. and A. Hertz: "Ants Can Colour Graphs", Journal of the Operational Research Society, 48, 295-305 (1997).
118. Kuntz P., P. Layzell and D. Snyers: "A Colony of Ant-like Agents for Partitioning in VLSI Technology", Proceedings of the Fourth European Conference on Artificial Life, P. Husbands and I. Harvey, (Eds.), 417-424, MIT Press (1997).
119. Schoonderwoerd R., O. Holland, J. Bruten and L. Rothkrantz: "Ant-based Load Balancing in Telecommunications Networks", Adaptive Behavior, 5(2):169–207 (1997).

120. Navarro Varela G. and M.C. Sinclair: "Ant Colony Optimisation for Virtual-Wavelength-Path Routing and Wavelength Allocation", Proceedings of the Congress on Evolutionary Computation (CEC'99), Washington DC, USA, July (1999).
121. Yanowski, V. Wagner I.A., and Bruckstein A.M., "A Distributed Ant Algorithm for Efficiently Patrolling a Network", Workshop on Interdisciplinary Applications of Graph Theory and Algorithms, Haifa, Israel, April 17-18, (2001).
122. Machado A., Ramalho G., Zucker J.D., Drogoul A.: "Multi-Agent Patrolling: an Empirical Analysis of Alternative Architectures", Proceedings of MABS'02 (Multi-Agent Based Simulation, Bologna, Italy, July 2002), LNCS, Springer-Verlag (2002).
123. Rouff C., Hinchey M., Truszkowski W., Rash J.: "Verifying large numbers of cooperating adaptive agents", Parallel and Distributed Systems, Proceedings of the 11th International Conference on Volume 1, 20-22 July 2005 Page(s):391 - 397 Vol. 1 (2005).
124. P. Scerri, E. Liao, Yang. Xu, M. Lewis, G. Lai, and K. Sycara: "Coordinating very large groups of wide area search munitions", Theory and Al gorithms for Cooperative Systems, chapter. World Scientific Publishing, (2004).
125. V. Vazirani: "Approximation Algorithms", Springer-Verlag, (2001).
126. S. Arora, S. Safra: "Probabilistic checking of proofs: A new characterization of NP", Journal of the ACM, (1998).
127. M. Garey, D. Johnson: "Computers and Intractability: A Guide to the Theory of NP-Completeness", San Francisco, CA: W. H. Freeman, (1979).
128. Y. Altshuler, A. Matsliah, A. Felner: "On the Complexity of Physical Problems and a Swarm Algorithm for k-Clique Search in Physical Graphs", European Conference on Complex Systems (ECCS-05), Paris, France, November (2005).
129. A.Z. Broder, A.R. Karlin, P. Raghavan, E. Upfal: "Trading Space for Time in Undirected $s - t$ Connectivity", ACM Symposium on Theory of Computing (STOC), pp. 543–549, (1989).

6

Ant Colony Optimisation for Fast Modular Exponentiation using the Sliding Window Method

Nadia Nedjah[1] and Luiza de Macedo Mourelle[2]

[1] Department of Electronics Engineering and Telecommunications,
   Faculty of Engineering, State University of Rio de Janeiro.
   nadia@eng.uerj.br, http://www.eng.uerj.br/~nadia/english.html
[2] Department of Systems Engineering and Computation,
   Faculty of Engineering, State University of Rio de Janeiro.
   ldmm@eng.uerj.br, http://www.eng.uerj.br/~ldmm

Modular exponentiation is the main operation to RSA-based public-key cryptosystems. It is performed using successive modular multiplications. This operation is time consuming for large operands, which is always the case in cryptography. For software or hardware fast cryptosystems, one needs thus reducing the total number of modular multiplications required. Existing methods attempt to reduce this number by partitioning the exponent in constant or variable size windows. However, these window-based methods require some pre-computations, which themselves consist of modular exponentiations. It is clear that pre-processing needs to be performed efficiently also. In this chapter, we exploit the ant colony strategy to finding an optimal addition sequence that allows one to perform the pre-computations in window-based methods with a minimal number of modular multiplications. Hence we improve the efficiency of modular exponentiation. We compare the yielded addition sequences with those obtained using Brun's algorithm.

6.1 Introduction

Public-key cryptographic systems (such as the RSA encryption scheme [6], [12]) often involve raising large elements of some groups fields (such as $GF(2^n)$ or elliptic curves [9]) to large powers. The performance and practicality of such cryptosystems is primarily determined by the implementation efficiency of the modular exponentiation. As the operands (the plain text of a message or the cipher (possibly a partially ciphered) are usually large (i.e. 1024 bits or more),

N. Nedjah and Luiza de Macedo Mourelle: *Ant Colony Optimisation for Fast Modular Exponentiation using the Sliding Window Method*, Studies in Computational Intelligence (SCI)
26, 133–147 (2006)
www.springerlink.com   © Springer-Verlag Berlin Heidelberg 2006

and in order to improve time requirements of the encryption/decryption operations, it is essential to attempt to minimise the number of modular multiplications performed.

A simple procedure to compute $C = T^E \bmod M$ based on the paper-and-pencil method is described in Algorithm 1. This method requires E-1 modular multiplications. It computes all powers of $T : T \rightarrow T^2 \rightarrow \cdots \rightarrow T^{E-1} \rightarrow T^E$.

Algorithm 1. simpleExponentiationMethod(T, M, E)
1. $C := T$;
2.   for $i := 1$ to $E - 1$ do $C := (C \times T) \bmod M$;
3.   return C;
end algorithm.

The computation of exponentiations using Algorithm 1 is very inefficient. The problem of yielding the power of a number using a minimal number of multiplications is NP-hard [5], [10]. There are several efficient algorithms that perform exponentiation with a nearly minimal number of modular multiplications, such that the window-based methods. However, these methods need some pre-computations that if not performed efficiently can deteriorate the algorithm overall performance. The pre-computations are themselves an ensemble of exponentiations and so it is also NP-hard to perform them optimally.

In this chapter, we concentrate on this problem and engineer a new way to do the necessary pre-computations very efficiently. We do so using the ant colony methodology. We compare our results with those obtained using the Brun's algorithm [1].

Ant systems [2-1] are distributed multi-agent systems [3-1] that simulate real ant colony. Each agent behaves as an ant within its colony. Despite the fact that ants have very bad vision, they always are capable to find the shortest path from their nest to wherever the food is. To do so, ants deposit a trail of a chemical substance called *pheromone* on the path they use to reach the food. On intersection points, ants tend to choose a path with high amount of pheromone. Clearly, the ants that travel through the shorter path are capable to return quicker and so the pheremone deposited on that path increases relatively faster than that deposited on much longer alternative paths. Consequently, all the ants of the colony end using the shorter way.

In this chapter, we exploit the ant colony methodology to obtain an optimal solution to AS-chain minimisation NP-complete problem. In order to clearly report the research work performed, we subdivide the rest of this chapter into five important sections. In Section 6.2, we present the window methods; In Section 6.3, we present the concepts of addition chains and sequence and they can be used to improve the pre-computations of the window methods; In Section 6.4, we give an overview on the concepts of ant colony optimisation ; In Section 6.5, we explain how these concepts can be used to compute a minimal addition chain to perform efficiently necessary pre-computations in the window methods. In Section 6.6, we present some useful results.

6.2 Window-Based Methods

Generally speaking, the window methods for exponentiation [5] may be thought of as a three major step procedure:

i. partitioning in k-bits windows the binary representation of the exponent E;
ii. pre-computing the powers in each window one by one;
iii. iterating the squaring of the partial result k times to shift it over, and then multiplying it by the power in the next window when if window is not 0.

There are several partitioning strategies. The window size may be constant or variable. For the m-ary methods, the window size is constant and the windows are next to each other. On the other hand, for the sliding window methods the window size may be of variable length. It is clear that zero-windows, i.e. those that contain only zeros, do not introduce any extra computation. So a good strategy for the sliding window methods is one that attempts to maximise the number of zero-windows. The details of m-ary methods are exposed in Section 6.2.1 while those related to sliding constant-size window methods are given in Section 6.2.2. In Section 6.2.3, we introduce the adaptive variable-size window methods.

6.2.1 *M*-ary Methods

The m-ary methods [3] scans the digits of E form the less significant to the most significant digit and groups them into partitions of equal length $\log_2 m$, where m is a power of two. Note that 1-ary methods coincides with the square-and- multiply well-known binary exponentiation method.

In general, the exponent E is partitioned into p partitions, each one containing $l = \log_2 m$ successive digits. The ordered set of the partition of E will be denoted by $\mathbb{P}(E)$. If the last partition has less digits than $\log_2 m$, then the exponent is expanded to the left with at most $log_2 m - 1$ zeros. The m-ary algorithm is described in Algorithm 2, wherein V_i denotes the decimal value of partition P_i.

Algorithm 2. m-aryMethod(T, M, E)
1.   Partition E into p l-digits partitions;
2.   for $i := 2$ to m Compute $T^i \bmod M$;
3. $C := T^{V_p} \bmod M$;
4.   for $i := p - 2$ downto 0

5. $C := C^{2_l} \bmod M$;
6.   if $V_i \neq 0$ then $C := C\times \bmod M$;
7.   return C;
end algorithm.

6.2.2 Sliding Window Methods

For the sliding window methods the window size may be of variable length and hence the partitioning may be performed so that the number of zero-windows is as large as possible, thus reducing the number of modular multiplication necessary in the squaring and multiplication phases. Furthermore, as all possible partitions have to start (i.e. in the right side) with digit 1, the pre-processing step needs to be performed for odd values only. The sliding method algorithm is presented in Algorithm 3, wherein d denotes the number of digits in the largest possible partition and L_i the length of partition P_i.

Algorithm 3. slidingWindowMethod(T, M, E)
1.   Partition E using the given strategy;
2.   for $i := 2$ to $2^d - 1$ step 2 do Compute $T^i \bmod M$;
3. $C := T^{V_{p-1}} \bmod M$;
4.   for $i := p - 2$ downto 0 do
5. $C := C^{L_i} \bmod M$;
6.   if $V_i \neq 0$ then $C := C \times T^{V_i} \bmod M$;
7.   return C;
end algorithm.

In adaptive methods [7] the computation depends on the input data, such as the exponent E. M-ary methods and window methods pre-compute powers of all possible partitions, not taking into account that the partitions of the actual exponent may or may not include all possible partitions. Thus, the number of modular multiplications in the pre-processing step can be reduced if partitions of E do not contain all possible ones.

Let $\wp(E)$ be the list of partitions obtained from the binary representation of E. Assume that the list of partition is non-redundant and ordered according to the ascending decimal value of the partitions contained in the expansion of E. Recall that V_i and L_i are the decimal value and the number of digits of partition P_i. The generic algorithm for describing the computation of $T^E \bmod M$ using the window methods is given in Algorithm 4.

Algorithm 4. AdaptiveWindowMethod(T, M, E)
1.   Partition E using the given strategy;
2.   for each partition $P_i \in \wp$ do Compute $T^{V_i} \bmod M$;
3. $C := T^{V_{p-1}} \bmod M$;
4.   for $i := p - 2$ downto 0 do
5. $C := C^{L_i} \bmod M$;

6.   if $V_i \neq 0$ then $C := C \times T^{V_i} \bmod M$;
7.   return C;
end algorithm.

In Algorithm 2 and Algorithm 3, it is clear how to perform the pre-computation indicated in Line 2. For instance, let $E = 1011001101111000$. The pre-processing step of the 4-ary method needs 14 modular multiplications $(T \to T \times T = T^2 \to T \times T^2 = T^3 \to \ \to T \times T^{14} = T^{15})$ and that of the maximum 4-digit sliding window method needs only 8 modular multiplications $(T \to T \times T = T^2 \to T \times T^2 = T^3 \to T^3 \times T^2 = T^5 \to T^5 \times T^2 = T^7 \to \ \to T^{13} \times T^2 = T^{15})$. However the adaptive 4-ary method would partition the exponent as $E = 1011\|0011\|0111\|1000$ and hence needs to pre-compute the powers T^3, T^7, T^8 and T^{11} while the method maximum 4-digit sliding window method would partition the exponent as $E = 1\|0\|11\|00\|11\|0\|1111\|000$ and therefore needs to pre-compute the powers T^3 and T^{15}. The pre-computation of the powers needed by the adaptive 4-digit sliding window method may be done using 6 modular multiplications $T \to T \times T = T^2 \to T \times T^2 = T^3 \to T^2 \times T^2 = T^4 \to T^3 \times T^4 = T^7 \to T^7 \times T = T^8 \to T^8 \times T^3 = T^{11}$ while the pre-computation of those powers necessary to apply the adaptive sliding window may be accomplished using 5 modular multiplications $T \to T \times T = T^2 \to T \times T^2 = T^3 \to T^2 \times T^3 = T^5 \to T^5 \times T^5 = T^{10} \to T^5 \times T^{10} = T^{15}$. Note that Algorithm 4 does not suggest how to compute the powers (Line 2) needed to use the adaptive window methods. Finding the best way to compute them is a NP-hard problem [4], [7].

6.3 Addition Chains and Addition Sequences

An *addition chain* of length l for an positive integer N is a list of positive integers $(E_1, E_2, \ldots, E_l)$ such that $E_1 = 1$, $E_l = N$ and $E_k = E_i + E_j$, $0 \leq i \leq j < k \leq l$. Finding a minimal addition chain for a given positive integer is an NP-hard problem. It is clear that a short addition chain for exponent E gives a fast algorithm to compute $T^E \bmod M$ as we have if $E_k = E_i + E_j$ then $T^{E_k} = T^{E_i} \times T^{E_j}$. The adaptive window methods described earlier use a near optimal addition chain to compute $T^E \bmod M$. However these methods do not prescribe how to perform the pre-processing step (Line 3 of Algorithm 4). In the following we show how to perform this step with minimal number of modular multiplications.

6.3.1 Addition Sequences

There is a generalisation of the concept of addition chains, which can be used to formalise the problem of finding a minimal sequence of powers that should be computed in the pre-processing step of the adaptive window method.

An *addition sequence* for the list of positive integers $V_1, V_2, \ldots, V_p$ such that $V_1 < V_2 < \cdots < V_p$ is an addition chain for integer V_p that includes all the integers $V_1, V_2, \ldots, V_p$. The length of an addition sequence is the numbers of integers that constitute the chain. An addition sequence for a list of positive integers $V_1, V_2, \ldots, V_p$ will be denoted by $\xi(V_1, V_2, \ldots, V_p)$.

Hence, to optimise the number of modular required multiplications in the pre-processing step of the adaptive window methods for computing T^E mod M, we need to find an addition sequence of minimal length (or simply minimal addition sequence) for the values of the partitions included in the non-redundant ordered list $\wp(E)$. This is an NP-hard problem and we use genetic algorithm to solve it. Our method showed to be very effective for large window size. General principles of genetic algorithms are explained in the next section.

6.3.2 Brun's Algorithm

Now we describe briefly, Brun's algorithm [1] to compute addition sequences. The algorithm is a generalisation of the continued fraction algorithm [1]. Assume that we need to compute the addition sequence $\xi(V_1, V_2, \ldots, V_p)$. Let $Q = \lfloor \frac{V_p}{V_{p-1}} \rfloor$ and let $\chi(Q)$ be the addition chain for Q using the binary method (i.e. that used in Algorithm 2 with $l = 1$). Let $R = V_p - Q \times V_{p-1}$. By induction we can construct an addition sequence $\xi(V_1, V_2, \ldots, R, \ldots, V_{p-1})$, then obtain:

$$\xi(V_1, V_2, \ldots, V_p) = \xi(V_1, V_2, \ldots, R, \ldots, V_{p-1}) \cup \\ V_{p-1} \times \chi(Q) \setminus \{1\} \cup \{V_p\} \tag{6.1}$$

6.4 Ant Systems and Algorithms

Ant systems can be viewed as multi-agent systems [3] that use a shared memory through which they communicate and a local memory to bookkeep the locally reached problem solution. Fig. 6.1. depicts the overall structure of an system, wherein A_i and LM_i represent the $i^{th.}$ agent of the ant system and its local memory respectively. Mainly, the shared memory (SM) holds the pheromone information while the local memory LM_i keeps the solution (possibly partial) that agent A_i reached so far.

The behaviour of an artificial ant colony is summarised in Algorithm 4, wherein N, C, SM are the number of of artificial ant that form the colony, the characteristics of the expected solution and the shared memory used by the artificial ants to store pheromone information repsectively. The first step consists of activating N distinct artificial ants that should work in simultaneously. Every time an ant conclude its search, the shared memory is updated with an amount of pheromone, which should be proportional to the quality of the reached solution. This called *global* pheromone update. When the solution

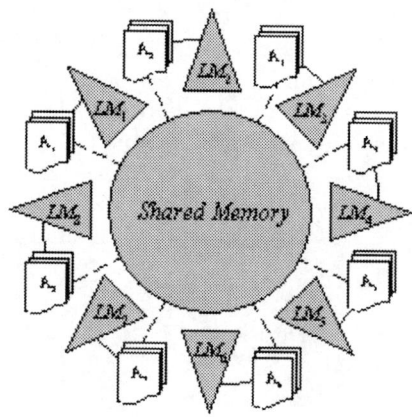

Fig. 6.1. Multi-agent system architecture

yield by an ant's work is suitable (i.e. fits characteristc C) then all the active ants are stopped. Otherwise, the process is iterated until an adequate solution is encountered.

Algorithm 4. ArtificialAntColony(N, C)
1:    Initialise SM with initial pheromone;
2:    do
3:        for $i := 1$ to N
4:            Start ArtificialAnt(A_i, LM_i);
5:            $Active := Active \cup \{A_i\}$;
6:        do
7:            Update SM w.r.t. pheromone evaporation;
8:            when an ant (say A_i) halts do
9:                $Active := Active \setminus \{A_i\}$;
10:               $\Phi :=$ Pheromone(LM_i);
11:               Update SM with global pheromone Ψ;
12:               $S :=$ ExtractSolution(LM_i);
13:           until Characteristics(S) $= C$ or $Active = \emptyset$;
14:           while $Active \neq \emptyset$ do
15:               Stop ant $A_i \mid A_i \in Active$;
16:               $Active := Active \setminus \{A_i\}$;
17:       until Characteristics(S) $= C$;
18:       return S;
end.

The behaviour of an artificial ant is described in Algorithm 5, wherein A_i and LM_i represent the ant identifier and the ant local memory, in which it stores the solution computed so far. First, the ant computes the probabilities

that it uses to select the next state to move to. The computation depends on
the solution built so far, the problem constraints as well as some heuristics
[2], [6]. Thereafter, the ant updates the solution stored in its local memory,
deposits some *local* pheromone into the shared memory then moves to the chosen state. This process is iterated until complete problem solution is yielded.

Algorithm 5. ArtificialAnt(A_i, LM_i)
1:    Initialise LM_i;
2:    do
3:        $P :=$ TransitionProbabilities(LM_i);
4:        $NextState :=$ StateDecision(LM_i, P);
5:        Update LM_i; Update SM with local pheromone;
6:        $CurrentState := NextState$);
7:    until $CurrentState := TargetState$;
8:    Halt A_i;
end.

6.5 Chain Sequence Minimisation Using Ant System

In this section, we concentrate on the specialisation of the ant system of Algorithm 4 and Algorithm 5 to the addition sequence minimisation problem. For
this purpose, we describe how the shared and local memories are represented.
We then detail the function that yields the solution (possibly partial) characteristics. Thereafter, we define the amount of pheromone to be deposited
with respect to the solution obtained so far. Finally, we show how to compute
the necessary probabilities and make the adequate decision towards a shorter
addition sequence for the considered the sequence $(V_1, V_2, \ldots, V_p)$.

6.5.1 The Ant System Shared Memory

The ant system shared memory is a two-dimension array. If the last exponent
in the sequence is V_p then the array should V_p rows. The number of columns
depends on the row. It can be computed as in (6.2), wherein NC_i denotes the
number of columns in row i.

$$NC_i = \begin{cases} 2^{i-1} - i + 1 & \text{if } 2^{i-1} < V_p \\ 1 & \text{if } i = V_p \\ V_p - i + 3 & \text{otherwise} \end{cases} \quad (6.2)$$

An entry $SM_{i,j}$ of the shared memory holds the pheromone deposited by ants
that used exponent $i + j$ as the i th. member in the built addition sequence.
Note that $1 \le i \le V_p$ and for row i, $0 \le j \le NC_i$. Fig. 6.2 gives an example

of the shared memory for exponent 17. In this example, a table entry is set to show the exponent corresponding to it. The exponent $E_{i,j}$ corresponding to entry $SM_{i,j}$ should be obtainable from exponents of previous rows. Equation (6.3) formalises such a requirement.

$$E_{i,j} = E_{k_1,l_1} + E_{k_2,k_2} \mid \begin{array}{l} 1 \le k_1, k_2 < i, 0 \le l_1, l_2 \le j, \\ k_1 = k_2 \iff l_1 = l_2 \end{array} \tag{6.3}$$

1											
2											
3	4										
4	5	6	7	8							
5	6	7	8	9	10	11	12	13	14	15	16
6	7	8	9	10	11	12	13	14	15	16	17
7	8	9	10	11	12	13	14	15	16	17	
8	9	10	11	12	13	14	15	16	17		
9	10	11	12	13	14	15	16	17			
10	11	12	13	14	15	16	17				
11	12	13	14	15	16	17					
12	13	14	15	16	17						
13	14	15	16	17							
14	15	16	17								
15	16	17									
16	17										
17											

Fig. 6.2. Example of shared memory content for $V_p = 17$

Note that, in Fig. 6.2, the exponents in the shaded entries are not valid exponents as for instance exponent 7 of row 4 can is not obtainable from the sum of two previous different stages, as described in (6.3). The computational process that allows us to avoid these exponents is of very high cost. In order to avoid using these few exponents, we will penalise those ants that use them and hopefully, the solutions built by the ants will be almost all valid addition chains. Furthermore, note that for a valid solution need also to contain all the exponents of the sequence i.e., $V_1, V_2, \ldots, V_{p-1}, V_p$.

6.5.2 The Ant Local Memory

In an ant system, each ant is endowed a local memory that allows it to store the solution or the part of it that was built so far. This local memory is divided into two parts: the first part represents the (partial) addition sequence found by the ant so far and consists of a one-dimension array of V_p entries; the second part holds the *characteristic* of the solution. It represents the solution fitness i.e., its length. The details of how to compute the fitness of a possibly partial addition sequence are given in the next section. Fig. 6.3 shows six different examples of an ant local memory for sequence (5, 7, 11). Fig. 6.3(a) represents addition sequence (1, 2, 4, 5, 7, 11), which is a valid and complete solution of fitness 5. Fig. 6.3(b) depicts addition sequence (1, 2, 3, 5, 7, 10, 11), which is also a valid and complete solution but of fitness 6. Fig. 6.3(c) represents partial addition sequence (1, 2, 4, 5), which is a valid and but incomplete solution as it does not include exponent 7 and 11 and the last exponent is smaller than both 7 and 11. The corresponding fitness is 8.8. Fig. 6.3(d) consists of non-valid addition sequence (1, 2, 4, 5, 10, 11) as 7 is not included. The corresponding fitness is 15. Fig. 6.3(e) represents also non-valid addition sequence (1, 2, 3, 5, 7, 11) as 11 is not a sum two previous exponents in the sequence. Its fitness is also 15. Finally, Fig. 6.3(f) represents also non-valid addition sequence (1, 2, 5, 10, 11) as 5 is not a sum two previous and mandatory exponent 7 is not in the addition sequence. exponents in the sequence. Its fitness is also 25. In next section, we explain how the fitness of a solution is computed.

6.5.3 Addition Sequence Characteristics

The fitness evaluation of an addition sequence is performed with respect to three aspects: *(a)* how much it adheres to the definition (see Section 6.3), i.e. how many of its members cannot be obtained summing up two previous members of the sequence; *(b)* how far the it is reduced, i.e. what is the length of the chain; *(c)* how many of the mandatory exponents do not appear in the sequence. (6.4) shows how to compute the fitness f of solution $(E_1, E_2, \ldots, E_n, 0, \ldots, 0)$ regarding mandatory exponents $V_1, V_2, \ldots, V_p$.

$$f(V_1, V_2, \ldots, V_p, E_1, E_2, \ldots, E_n) = \frac{V_p \times (n-1)}{E_n} + (\eta_1 + \eta_2) \times penalty \tag{6.4}$$

wherein η_1 represents the number of E_i, $3 \leq i \leq n$ in the addition sequence that verify the predicate below:

$$\forall j, k \mid 1 \leq j, k < i, E_i \neq E_j + E_k \tag{6.5}$$

and $\eta 2$ represents the number of mandatory exponents V_i, $1 \leq i \leq p$ that verify the predicate below:

$$V_i \leq E_n \implies \forall j \mid 1 \leq j \leq n, E_j \neq V_i \tag{6.6}$$

For a valid complete addition sequence, the fitness coincides with its length, which is the number of multiplications that are required to compute the exponentiation using the sequence. For a valid but incomplete addition

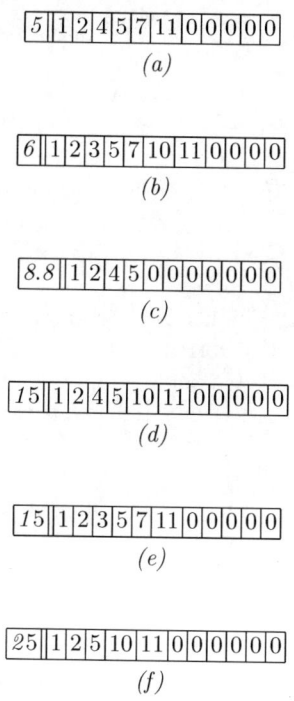

| 5 | 1 | 2 | 4 | 5 | 7 | 11 | 0 | 0 | 0 | 0 |

(a)

| 6 | 1 | 2 | 3 | 5 | 7 | 10 | 11 | 0 | 0 | 0 | 0 |

(b)

| 8.8 | 1 | 2 | 4 | 5 | 0 | 0 | 0 | 0 | 0 | 0 | 0 |

(c)

| 15 | 1 | 2 | 4 | 5 | 10 | 11 | 0 | 0 | 0 | 0 | 0 |

(d)

| 15 | 1 | 2 | 3 | 5 | 7 | 11 | 0 | 0 | 0 | 0 | 0 |

(e)

| 25 | 1 | 2 | 5 | 10 | 11 | 0 | 0 | 0 | 0 | 0 | 0 |

(f)

Fig. 6.3. Examples of an ant local memory: *(a)* complete valid addition sequence of fitness 5; *(b)* complete valid addition sequence of fitness 6; *(c)* incomplete valid addition sequence of fitness 8.8; *(d)*, *(e)* complete non-valid solution of fitness 15; *(f)* complete non-valid addition sequence of fitness 25

sequence, the fitness consists of its *relative* length. It takes into account the distance between last mandatory exponent V_p and the last exponent in the partial addition sequence. Furthermore, for every mandatory exponent that is smaller than the last member of the sequence which is not part of it, a penalty is added to the sequence fitness. Note that valid incomplete sequences may have the same fitness of some other valid and complete ones. For instance, addition sequence (1, 2, 3, 6, 8) and (1, 2, 3, 6) for exponent mandatory exponents (3, 6, 8) have the same fitness 4.

For an invalid addition sequences, a penaly, which should be larger than V_p, is introduced into the fitness value for each exponent for which one cannot find two (may be equal) members of the sequence whose sum is equal to the exponent in question or two distincts previous members of the chain whose difference is equal to the considered exponent. Furthermore, a penalty is added to the fitness of a addition sequence whenever the a mandatory exponent is not part of it. The penalty used in the examples of Fig. 6.3 is 10.

6.5.4 Pheromone Trail and State Transition Function

There are three situations wherein the pheromone trail is updated: *(a)* when an ant chooses to use exponent $F = i + j$ as the *i*th. member in its solution, the shared momory cell $SM_{i,j}$ is incremented with a constant value of pheromone $\Delta\phi$, as in (6.7); *(b)* when an ant halts because it reached a complete solution, say $\alpha = (E_1, E_2, \ldots, E_n)$ for madatory exponent sequence σ, all the shared memory cells $SM_{i,j}$ such that $i + j = E_i$ are incremented with pheromone value of $1/Fitness(\sigma, \alpha)$, as in (6.8). Note that the better is the reached solution, the higher is the amount of pheromone deposited in the shared memory cells that correspond to the addition sequence members. *(iii)* The pheromone deposited should evaporate. Priodically, the pheromone amount stored in $SM_{i,j}$ is decremented in an exponential manner [6] as in (6.9).

$$SM_{i,j} := SM_{i,j} + \Delta\phi, \quad \text{every time } E_i = i + j \text{ is chosen} \tag{6.7}$$

$$SM_{i,j} := SM_{i,j} + 1/Fitness(\sigma, \alpha), \quad \forall i, j \mid i + j = E_i \tag{6.8}$$

$$SM_{i,j} := (1 - \rho)SM_{i,j} \mid \rho \in (0, 1], \quad \text{periodically} \tag{6.9}$$

An ant, say A that has constructed partial addition sequence $(E_1, E_2, \ldots, E_i, 0, 0)$ for exponent sequence $(V_1, V_2, \ldots, V_p)$, is said to be in *step i*. In step $i + 1$, it may choose exponent E_{i+1} $E_i + 1$, $E_i + 2$, $\ldots$, $2E_i$, if $2E_i \leq V_p$. That is, ant A may choose one of the exponents that are associated with the shared memory cells SM_{i+1,E_i-i}, SM_{i+1,E_i-i+1}, $\ldots$, $SM_{i+1,2E_i-i-1}$. Otherwise (i.e. if $2E_i > V_p$), it may only select from exponents $E_i + 1$, $E_i + 2$, $\ldots$, $E + 2$. In this case, ant A may choose one of the exponent associated with SM_{i+1,E_i-i}, SM_{i+1,E_i-i+1}, $\ldots$, $SM_{i+1,E-i+1}$. Furthermore, ant A chooses the new exponent E_{i+1} with the probability expressed through (6.10).

$$P_{i,j} = \begin{cases} \dfrac{SM_{i+1,j}}{\max_{k=E_i-i}^{2E_i-i-1} SM_{i+1,k}} & \text{if } 2E_i \leq E \ \& \\[2ex] & j \in [E_i - i, 2E_i - i - 1] \\[2ex] \dfrac{SM_{i+1,j}}{\max_{k=E_i-i}^{E-i-1} SM_{i+1,k}} & \text{if } 2E_i > E \ \& \\[2ex] & j \in [E_i - i, E - i - 1] \\[2ex] 0 & \text{otherwise} \end{cases} \tag{6.10}$$

6.6 Performance Comparison

The ant system described in Algorithm 3 was implemented using Java as a multi-threaded ant system. Each ant was simulated by a thread that implements the artificial ant computation of Algorithm 4. A Pentium IV-HT™ of a operation frequency of 1GH and RAM size of 2GB was used to run the ant system and obtain the performance results.

We compared the performance of m-ary methods, the Brun's algorithm, genetic algorithms and ant system-based methods. The obtained addition chains are given in Table 6.1 The average lengths of the addition sequences for different exponent sequences obtained using these methods are given in Table 6.2. The exponent size is that of its binary representation (i.e. number of bits). The ant system-based method always outperforms all the others, including the genetic algorithm-based method [7]. The chart of Fig. 6.4 shows the relation between the average length of the obtained addition sequences.

Table 6.1. The addition sequences yield for S(5, 9, 23), S(9, 27, 55) and S(5, 7, 95) respectively

Method	Addition sequence	#×
5-ary	(1,2,3,4,5,6,7,8,9,...,22,23,...,30,31)	30
5-window	(1,2,3,5,7,9,11,...,31)	16
Brun's	(1,2,4,5,9,18,23)	6
GAs	(1,2,4,5,9,18,23)	6
Ant system	(1,2,4,5,9,14,23)	6
6-ary	(1,2,3,...,8,9,...,26,27,...,54,55,...,63)	62
6-window	(1,2,3,...,7,9,...,25,27,...,53,55,...,63)	31
Brun's	(1,2,3,6,9,18,27,54,55)	8
GAs	(1,2,4,8,9,18,27,28,55)	8
Ant system	(1,2,4,5,9,18,27,54,55)	8
7-ary	(1,2,3,4,5,6,7,...,95)	94
7-window	(1,2,3,5,7,...,95)	43
Brun's	(1,2,4,5,7,14,21,42,84,91,95)	10
GAs	(1,2,3,5,7,10,20,30,35,65,95)	10
Ant system	(1,2,4,5,7,14,19,38,76,95)	9
7-ary	(1,2,3,4,5,6,7,...,95)	94
7-window	(1,2,3,5,7,...,95)	43
Brun's	(1,2,4,5,7,14,21,42,84,91,95)	10
GAs	(1,2,3,5,7,10,20,30,35,65,95)	10
Ant system	(1,2,4,5,7,14,19,38,76,95)	9

6.7 Summary

In this chapter we applied the methodology of ant colony to the addition chain minimisation problem. Namely, we described how the shared and local memories are represented. We detailed the function that computes the solution fitness. We defined the amount of pheromone to be deposited with respect to the solution obtained by an ant. We showed how to compute the necessary

Table 6.2. Average length of addition sequence for Brun's algorithm, genetic algorithms (GA) and ant system (AS) based methods

size of V_p m-ary	Brun's	GA	AS
32	41	42	45
64	84	85	86
128	169	170	168
256	340	341	331
512	681	682	658
1024	1364	1365	1313

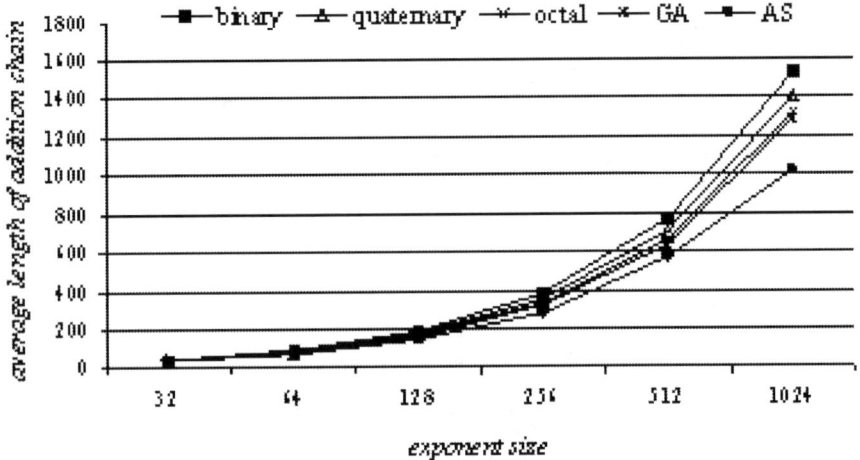

Fig. 6.4. Comparison of the average length of the addition chains for the binary, quaternary and octal methods vs. genetic algorithms and ant system-based methods

probabilities and make the adequate decision towards a good addition chain for the considered exponent. We implemented the ant system described using muti-threading (each ant of the system was implemented by a thread). We compared the results obtained by the ant system to those of m-ary methods (binary, quaternary and octal methods). Taking advantage of the a previous work on evolving minimal addition chains with genetic algorithm, we also compared the obtained results to those obtained by the genetic algorithm. The ant system always finds a shorter addition chain and gain increases with the size of the exponents.

References

1. Rivest, R., Shamir, A. and Adleman, L., A method for Obtaining Digital Signature and Public-Key Cryptosystems, Communications of the ACM, 21:120-126, 1978.
2. Dorigo, M. and Gambardella, L.M., Ant Colony: a Cooperative Learning Approach to the Travelling Salesman Problem, IEEE Transaction on Evolutionary Computation, Vol. 1, No. 1, pp. 53-66, 1997.
3. Feber, J., Multi-Agent Systems: an Introduction to Distributed Artificial Intelligence, Addison-Wesley, 1995.
4. Downing, P. Leong B. and Sthi, R., Computing Sequences with Addition Chains, SIAM Journal on Computing, vol. 10, No. 3, pp. 638-646, 1981.
5. Nedjah, N., Mourelle, L.M., Efficient Parallel Modular Exponentiation Algorithm, Second International Conference on Information systems, ADVIS'2002, Izmir, Turkey, Lecture Notes in Computer Science, Springer-Verlag, vol. 2457, pp. 405-414, 2002.
6. Stutzle, T. and Dorigo, M., ACO Algorithms for the Travelling Salesman Problems, Evolutionary Algorithms in Engineering and Computer Science, John-Wiley & Sons, 1999.
7. Nedjah, N. and Mourelle, L.M., Minimal addition-subtraction chains using genetic algorithms, Proceedings of the Second International Conference on Information Systems, Izmir, Turkey, Lecture Notes in Computer Science, Springer-Verlag, vol. 2457, pp. 303-313, 2002.

7

Particle Swarm for Fuzzy Models Identification

Arun Khosla[1], Shakti Kumar[2], K.K. Aggarwal[3], and Jagatpreet Singh[4]

[1] National Institute of Technology, Jalandhar – 144011, India.
khoslaak@nitj.ac.in
[2] Haryana Engineering College, Jagadhari – 135003, India.
[3] GGS Indraprastha University, Delhi – 110006, India. kka@ipu.edu
[4] Infosys Technologies Limited, Chennai – 600019, India. jagatpreet@yahoo.com

Fuzzy systems and evolutionary algorithms are two main constituents of computational intelligence paradigm and have their genesis in the nature-inspired extensions to the traditional techniques meant to solve problems of classification, control, prediction, modeling and optimization etc. Fuzzy systems are known for their capabilities to handle ambiguous or vague concepts of human perception for complex systems problems, where it is extremely difficult to describe the system models mathematically. On the other hand, the evolutionary algorithms have emerged as robust techniques for many complex optimization, identification, learning and adaptation problems. The objective of this chapter is to present the use of Particle Swarm Optimization (PSO) algorithm for building optimal fuzzy models from the available data. PSO, which is a robust stochastic evolutionary computation engine, belongs to the broad category of Swarm Intelligence (SI) techniques. SI paradigm has been inspired by the social behavior of ants, bees, wasps, birds, fishes and other biological creatures and is emerging as an innovative and powerful computational metaphor for solving complex problems in design, optimization, control, management, business and finance. SI may be defined as any attempt to design distributed problem-solving algorithms that emerges from the social interaction. The chapter also presents the results based on selection based PSO variant with lifetime parameter that has been used for identification of fuzzy models. The fuzzy model identification procedure using PSO as an optimization engine has been implemented as a Matlab toolbox viz. *PSO Fuzzy Modeler for Matlab* and is presented in the next chapter. The simulation results presented in this chapter have been obtained through this toolbox. The toolbox has been hosted on SourceForge.net, which is the world's largest development and download repository of open-source code and applications.

A. Khosla et al.: *Particle Swarm for Fuzzy Models Identification*, Studies in Computational Intelligence (SCI) **26**, 149–173 (2006)
www.springerlink.com

7.1 Introduction

Developing models of complex real-systems is an important topic in many disciplines of engineering. Models are generally used for simulation, identifying the system's behavior and design of controllers etc. Last few years have witnessed a drastic growth of sub-disciplines in science and engineering that have adopted the concepts of fuzzy set theory. This development can be attributed to successful applications in consumer electronics, robotics, signal processing, image processing, finance, management etc.

Design of fuzzy models or fuzzy model identification is the task of finding the parameters of fuzzy model so as to get the desired behavior. Two principally different approaches are used for the design of fuzzy models: heuristic-based design and model-based design. In the first approach, the design is constructed from the knowledge acquired from the expert, while in the second, the input-output data is used for building model. It is also possible to integrate both the approaches. In this chapter, we have presented the use of PSO algorithm for the identification of fuzzy models from the available data.

This chapter is organized into seven sections. In Section 7.2, a brief introduction to PSO algorithm is presented. Overview of fuzzy models alongwith various issues about fuzzy model identification problem are presented in Section 7.3. A methodology for fuzzy model identification using PSO algorithm is described in Section 7.4. This methodology has been implemented as a Matlab toolbox and the simulation results generated from this toolbox are reported in Section 7.5. In Section 7.6, a selection-based variant of PSO algorithm with a new defined parameter called lifetime is described briefly. This section also presents the simulation results based on this selection based new PSO variant that has been used for fuzzy models identification alongwith their analysis. Concluding remarks and some possible directions for future work are made in Section 7.7.

7.2 PSO Algorithm

The origin of PSO is best described as sociologically inspired, since it was initially developed as a tool by Reynolds [1][2] for simulating the flight patterns of birds, which was mainly governed by three major concerns: collision avoidance, velocity matching and flock centering. On the other hand, the reasons presented for the flocking behaviour observed in nature are: protection from predator and gaining from a large effective search with respect to food. The last reason assumes a great importance, when the food is unevenly distributed over a large region. It was realized by Kennedy and Eberhart that the bird flocking behavior can be adapted to be used as an optimizer and resulted in the first simple version of PSO [3] that has been recognized as one of the computational intelligence techniques intimately related to evolutionary algorithms. Like evolutionary computation techniques, it uses a population of potential solutions called particles that are flown through the hyperspace/search-space. In PSO, the particles have an adaptable velocity that determines their movement in the search-space. Each particle also has a memory and hence it is capable of remembering the best position in the search-space ever visited by it. The position corresponding to the best fitness is known as *pbest* and the overall best out of all the particles in the population is called *gbest*.

Consider that the search space is d-dimensional and i-th particle in the swarm can be represented by $X_i = (x_{i1}, x_{i2}, \ldots, x_{id})$ and its velocity can be represented by another d-dimensional vector $V_i = (v_{i1}, v_{i2}, \ldots, v_{id})$. Let the best previously visited position of this particle be denoted by $P_i = (p_{i1}, p_{i2}, \ldots, p_{id})$. If g-th particle is the best particle and the iteration number is denoted by the superscripts, then the swarm is modified according to (7.1) and (7.2) suggested by Shi and Eberhart [4]:

$$v_{id}^{n+1} = \chi(w v_{id}^{n} + c_1 r_1^{n} (p_{id}^{n} - x_{id}^{n}) + c_2 r_2^{n} (p_{gd}^{n} - x_{id}^{n})) \tag{7.1}$$

$$x_{id}^{n+1} = x_{id}^{n} + v_{id}^{n+1} \tag{7.2}$$

where,

$\chi-$ constriction factor
$w-$ inertia weight
c_1- cognitive acceleration parameter
c_2- social acceleration parameter
r_1, r_2- random numbers uniformly distributed in the range (0,1)

These parameters viz. inertia weight (w), cognitive acceleration (c_1), social acceleration (c_2), alongwith V_{max} [5] are known as the strategy/operating parameters of PSO algorithm. These parameters are defined by the user before the PSO run. The parameter V_{max} is the maximum velocity along any dimension, which implies that, if the velocity along any dimension exceeds V_{max}, it shall be clamped to this value. The inertia weight governs how much of the velocity should be retained from the previous time step. Generally the inertia weight is not kept fixed and is varied as the algorithm progresses so as to improve performance [4][5]. This setting allows the PSO to explore a large area at the start of simulation run and to refine the search later by a smaller inertia weight. The parameters c_1 and c_2 determine the relative pull of *pbest* and *gbest*. Random numbers r_1 and r_2 help in stochastically varying these pulls, that also account for slight unpredictable natural swarm behavior. Fig. 7.1 depicts the position update of a particle for a two-dimensional parameter space. Infact, this update is carried out as per (7.1) and (7.2) for each particle of swarm for each of the M dimensions in an M-dimensional optimization.

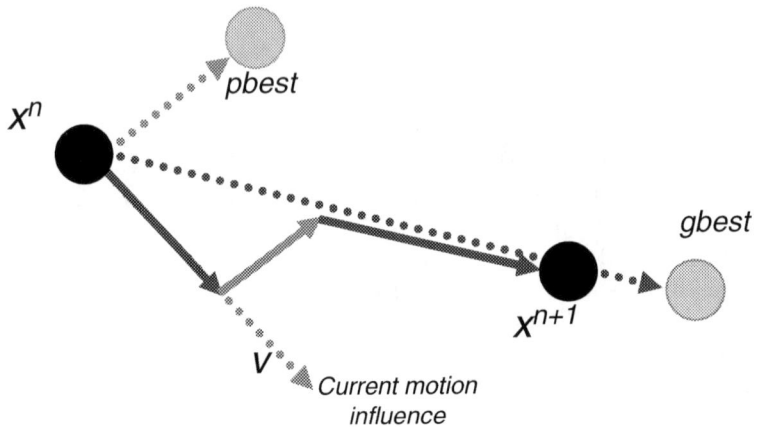

Fig. 7.1. Depiction of position updates in particle swarm optimization for 2-D parameter space

7.3 Fuzzy Models

This section reviews the fuzzy model structures and the various issues associated with the fuzzy model identification. Basic knowledge about the fuzzy logic, fuzzy sets and fuzzy inference system is assumed.

7.3.1 Overview of Fuzzy Models

Three commonly used types of fuzzy models are [6]:

- Mamdani-type fuzzy models
- Takagi-Sugeno fuzzy models
- Singleton fuzzy models

In Mamdani models, each fuzzy rule is of the form:
R_i: If x_1 is A_{i1} and ... and x_n is A_{in} then y is B

In Takagi-Sugeno models, each fuzzy rule is of the form:
R_i: If x_1 is A_{i1} and ... and x_n is A_{in} then y is $\sum_{i=1}^{n} a_i x_i + C$

whereas for Singleton model, each fuzzy rule is of the form:
R_i: If x_1 is A_{i1} and ... and x_n is A_{in} then y is C

where, $x_1, \ldots, x_n$ are the input variables and y is the output variable, $A_{i1}, \ldots, A_{in}$, B are the linguistic values of the input and output variables in the i-th fuzzy rule and C is a constant. Infact Singleton fuzzy model can seen as a special case of Takagi-Sugeno model, when $a_i = 0$. The input and output variables take their values in their respective universes of discourse or domains. Identification of Mamdani and Singleton fuzzy models has been considered in this chapter.

7.3.2 Fuzzy Model Identification Problem

Fuzzy modeling or fuzzy model identification is the task of identifying the parameters of fuzzy inference system so that a desired behaviour is achieved. Generally, the problem of fuzzy model identification includes the following issues [6][7]:

- Selecting the type of fuzzy model
- Selecting the input and output variables for the model
- Identifying the structure of the fuzzy model, which includes the determination of number and types of membership functions for the input and output variables and the number of fuzzy rules
- Identifying the parameters of antecedent and consequent membership functions
- Identifying the consequent parameters of the fuzzy rulebase

Some commonly used techniques for creating fuzzy models from the available input-output data are Genetic Algorithms [8][9][10][11][12], Fuzzy c-means (FCM) clustering algorithm [13][14], Neural Networks [6] and Adaptive Neuro Fuzzy Inference System model (ANFIS)[15][16].

7.4 A Methodology for Fuzzy Models Identification through PSO

Fuzzy model identification can be considered as an optimization process where part or all of the parameters of a fuzzy model constitute the search space. Each point in the search space corresponds to a fuzzy system i.e. represents membership functions, rule-base and hence the corresponding system behaviour. Given some objective/fitness function, the system performance forms a hypersurface and designing the optimal fuzzy system is equivalent to finding the optimal location on this hypersurface. The hypersurface is generally found to be infinitely large, nondifferentiable, complex, noisy, multimodal and deceptive [12], which make evolutionary algorithms very suitable for searching the hypersurface than the traditional gradient-based methods. PSO algorithms like GAs have the capability to find optimal or near optimal solution in a given complex search-space and can be used to modify/learn the parameters of fuzzy model. The methodology to identify the optimal fuzzy models using PSO as an optimization engine is shown in Fig. 7.2.

An optimization problem can be represented as a tuple of three components as represented in Fig. 7.3 and explained below:

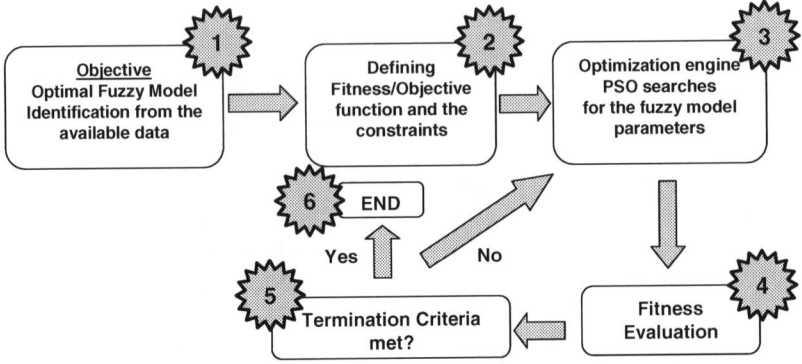

Fig. 7.2. Optimal fuzzy model identification using PSO as an optimization engine

- **Solution Space** – The first step in the optimization step is to pick up the variables to be optimized and define the domain/range in which to search for the optimal solutions.
- **Constraints** – It is required to define a set of constraints which must be followed by the solutions. Solutions not satisfying constraints are invalid solutions.
- **Fitness/Objective Function** – The fitness/objective function represents the quality of each solution and also provides a link between the optimization algorithm and the problem under consideration.

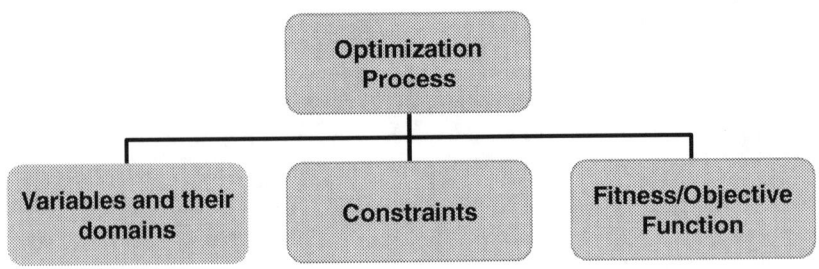

Fig. 7.3. Representation of optimization process

The objective of optimization problem is to look for the values of the variables being optimized, that satisfy the defined constraints, which maximizes or minimizes the fitness function. Hence, it is required to define the solution space, constraints and the fitness function when using PSO for the identification of optimal fuzzy models.

In this chapter, we have used Mean Square Error (MSE) defined in (7.3) as fitness/objective function for rating the fuzzy models.

$$MSE = \frac{1}{N} \sum_{k=1}^{N} [y(k) - \bar{y}(k)]^2 \qquad (7.3)$$

where,

$y(k)-$ desired output
$\bar{y}(k)-$ actual output of the model
$N-$ number of data points taken for model validation

A very important consideration is to completely represent a fuzzy system by a particle, and for this, all the needed information about the rule-base and membership functions is required to be specified through some encoding mechanism. It is also suggested to modify the membership functions and rule-base simultaneously, since they are codependent in a fuzzy system [12]. In this chapter, the methodology for identification of fuzzy models through PSO has been presented for three different cases, the details of which are provided in Table 7.1.

For the purpose of fuzzy model encoding, consider a multi-input single-output (MISO) system with n number of inputs. The number of fuzzy sets for the inputs are m_1, m_2, m_3,..., m_n respectively. In this chapter, we have considered only MISO fuzzy models, as multi-input multi-output (MIMO) models can be constructed by the parallel connection of several MISO models.

Table 7.1. Different Cases for Fuzzy Models Identification

Parameters modified through PSO	MF parameters	MF type	Rule consequents	Rule-set
Case I	Yes	No	Yes	No
Case II	Yes	Yes	Yes	No
Case III	Yes	Yes	Yes	Yes

7.4.1 Case I - *Parameters Modified*: MF parameters, rules consequents. *Parameters not Modified*: MF type, rule-set

Following assumptions have been made for encoding:

- Fixed numbers of triangular membership functions were used for both input and output variables with their centres fixed and placed symmetrically over corresponding universes of discourse.
- First and last membership functions of each input and output variable were represented with left- and right-skewed triangles respectively.

- Complete rule-base was considered, where all possible combinations of input membership functions of all the input variables were considered for rule formulation.
- Overlapping between the adjacent membership functions for all the variables was ensured through some defined constraints.

Encoding Mechanism (Membership functions)

Consider a triangular membership function and let parameters x_k^l, x_k^c and x_k^r represents the coordinates of left anchor, cortex and right anchor of kth linguistic variable as shown in the Fig. 7.4.

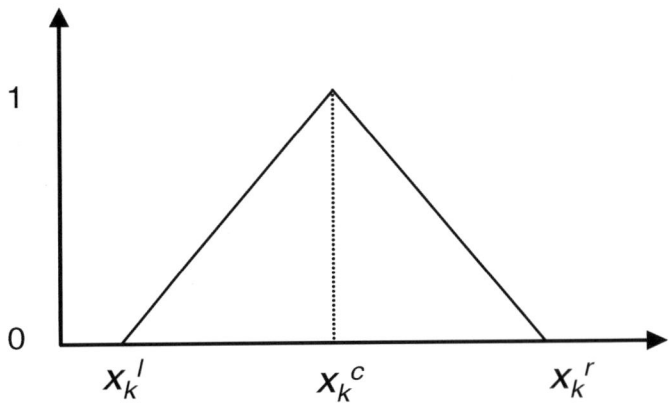

Fig. 7.4. Characteristics of a triangular membership function

A straightforward way to characterize this membership function is by means of 3-tuple (x_k^l, x_k^c, x_k^r). Therefore, particle carrying details about the parameters of the membership functions of all the input and output variables can be represented as follows:

$$(x_1^l, x_1^c, x_1^r, x_2^l, x_2^c, x_2^r \ldots\ldots, x_n^l, x_n^c, x_n^r, x_{n+1}^l, x_{n+1}^c, x_{n+1}^r)$$

The index $n+1$ is associated with the membership functions of the output variable.

It was ensured that following constraints are followed by every membership function of input and output variables.

$$x_k^l < x_k^c < x_k^r$$

At the same time, the overlapping between the adjacent membership functions was also ensured by defining some additional constraints. Let's assume that a variable is represented by three fuzzy sets as in Fig. 7.5, then those additional constraints to ensure overlapping can be represented by the following inequality.

$$x_{min} \leq x_2^l < x_1^r < x_3^l < x_2^r \leq x_{max}$$

where, x_{min} and x_{max} are the minimum and maximum values of the variable respectively.

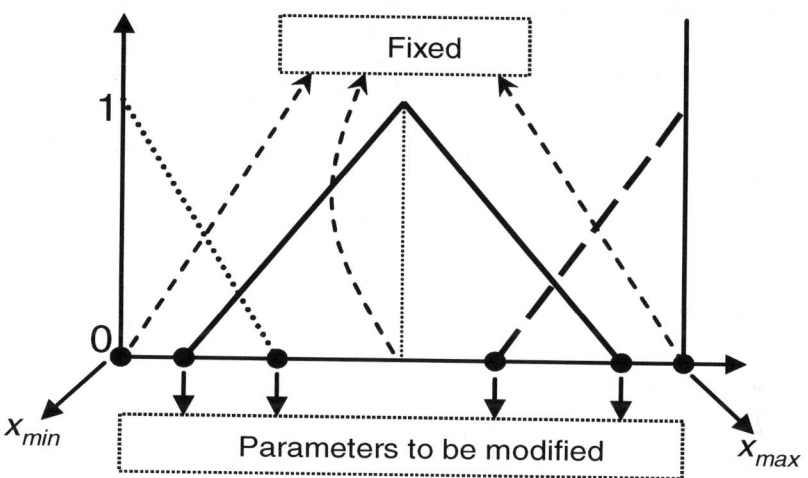

Fig. 7.5. Representation of a variable with 3 membership functions with centre of each membership function fixed and overlapping between the adjacent membership functions

The dimensions of the particle representing Mamdani fuzzy model can be worked out from Fig. 7.5, which represents the membership functions for any one of the input/output variables with three membership functions. Thus, four dimensions are required for each variable, which are to be modified during PSO run. The representation can be generalized to (7.4).

$$\texttt{Particle Size} = 2m_i - 2 \tag{7.4}$$

Thus the particle size for representing the membership functions of input and output variables for a Mamdani model is given by (7.5).

$$\text{Particle Size (for membership functions)} = \sum_{i=1}^{n+1}(2m_i - 2) \qquad (7.5)$$

where,

$n-$ number of input variables
m_i- number of fuzzy sets for i-th input and the index $n+1$ corresponds to the membership functions of the output variable.

Encoding Mechanism (Fuzzy Rules)

Considering the complete rule base, the particle size required for its representation is given by (7.6).

$$\text{Particle Size (for rule base)} = \prod_{i=1}^{n} m_i \qquad (7.6)$$

Thus, the particle size required for representing the complete Mamdani fuzzy model can be calculated through (7.7), obtained by adding (7.5) and (7.6).

$$\text{Particle Size (Mamdani Model)} = \sum_{i=1}^{n+1}(2m_i - 2) + \prod_{i=1}^{n} m_i \qquad (7.7)$$

If Singleton fuzzy model is considered with possible t number of consequent singleton values, then the particle dimensions required for representing this model can be obtained from (7.7) after a little modification and is represented by (7.8).

$$\text{Particle Size (Sugeno Model)} = \sum_{i=1}^{n}(2m_i - 2) + t + \prod_{i=1}^{n} m_i \qquad (7.8)$$

A particle representing a fuzzy model whose membership function parameters of input/output variables and rule consequents can be modified through PSO algorithm is shown in Fig. 7.6.

7.4.2 Case II - *Parameters Modified:* MF parameters, MFs type, rules consequents. *Parameters not Modified:* rule-set

The suggested methodology can be extended to increase the flexibility of search by incorporating additional parameters so as to execute the search for optimal solutions in terms of types of membership functions for each variable. Particle representing fuzzy model and implementing this approach is shown in Fig. 7.7.

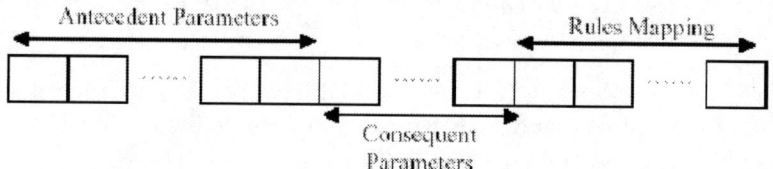

Fig. 7.6. Representation of a fuzzy model by a particle

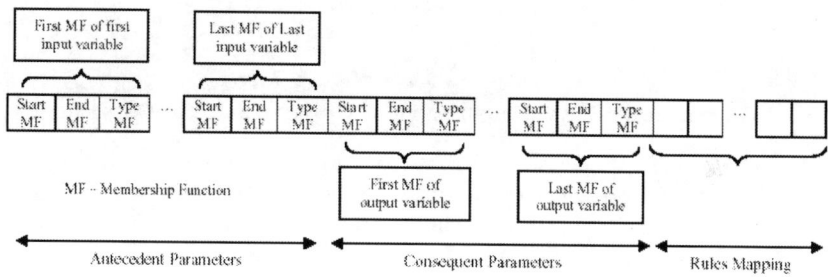

Fig. 7.7. Particle representing Mamdani fuzzy model corresponding to Case II, where MF parameters, MF types and rule consequents can be modified through PSO Algorithm

For such an implementation, the expression for particle size for encoding Mamdani fuzzy model would be as in (7.9).

$$\text{Particle Size (Mamdani Model)} = 3\sum_{i=1}^{n+1} m_i + \prod_{i=1}^{n} m_i \qquad (7.9)$$

The corresponding expression for the particle size to encode Singleton fuzzy model would be as given in (7.10).

$$\text{Particle Size (Sugeno Model)} = 3\sum_{i=1}^{n} 2m_i + t + \prod_{i=1}^{n} m_i \qquad (7.10)$$

In Fig. 7.7, each membership function is represented by three dimensions representing the start value, end value and the type of membership function like sigmodial, triangular etc. Like Case 1, complete rule-base has been considered here.

7.4.3 Case III - *Parameters Modified*: MF parameters, MFs type, rules consequents, rule-set

The methodology can be further extended so as to modify the rule-base also. For this implementation, the particle size for encoding Mamdani fuzzy model and Singleton fuzzy models would be as in (7.11) and (7.12).

$$\text{Particle Size (Mamdani Model)} = 3 \sum_{i=1}^{n+1} m_i + 2 \prod_{i=1}^{n} m_i \qquad (7.11)$$

$$\text{Particle Size (Sugeno Model)} = 3 \sum_{i=1}^{n} 2m_i + t + 2 \prod_{i=1}^{n} m_i \qquad (7.12)$$

Particle representing fuzzy model, where the MF parameters, MF type and rule-base can be modified through PSO is shown in Fig. 7.8. Two dimensions have been reserved for each rule, one representing the consequent value and other a flag. If the rule flag is '1', the rule is included, and for '0', it won't be part of the rule-base.

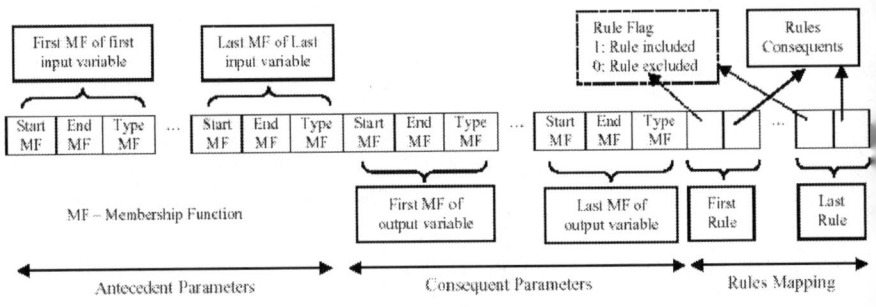

Fig. 7.8. Particle representing Mamdani fuzzy model corresponding to Case III, where MF parameters, MF types, rule consequents and rule-set can be modified through PSO Algorithm

Let's consider a system with two-inputs and single output. If we further consider that each input and output variable for this system is represented by three fuzzy sets, and five possible consequent values for the Singleton model, then the particle size for Mamdani and Singleton fuzzy models corresponding to Case II has been worked out as in (7.13) and (7.14) obtained from (7.9) and (7.10) respectively as below.

$$\text{Particle Size (Mamdani Model)} = 3\sum_{i=1}^{n+1} m_i + \prod_{i=1}^{n} m_i = 3*[3+3+3]+3*3 = 36$$

(7.13)

$$\text{Particle Size (Sugeno Model)} = 3\sum_{i=1}^{n} 2m_i + t + \prod_{i=1}^{n} m_i = 3*[3+3]+5+3*3 = 32$$

(7.14)

Similarly, the particle dimensions for the three cases considered can be calculated from the corresponding equations developed and are listed in Table 7.2.

Table 7.2. Particle size for three different cases defined in Table 7.1

Case/Model	Mamdani	Singleton
Case I	21	22
	Equation(7.7)	Equation(7.8)
Case II	36	32
	Equation(7.9)	Equation(7.10)
Case III	45	41
	Equation(7.11)	Equation(7.12)

The methodology for the identification of fuzzy model through PSO algorithm is represented as a flowchart in Figure 7.9.

7.5 Simulation Results

The proposed methodology has been applied for identification of fuzzy models for the rapid Nickel-Cadmium (Ni-Cd) battery charger, developed by the authors [17]. Based on the rigorous experimentation with the Ni-Cd batteries, it was observed that the two input variables used to control the charging rate (Ct) are absolute temperature of the batteries (T) and its temperature gradient (dT/dt). Charging rates are expressed as multiple of rated capacity of the battery, e.g. C/10 charging rate for a battery of C=500 mAh is 50 mA [18]. From the experiments performed, input-output data was tabulated and that data set consisting of 561 points is available at http://research.4t.com. The input and output variables identified for rapid Ni-Cd battery charger along with their universes of discourse are listed in Table 7.3.

The toolbox viz. *PSO Fuzzy Modeler for Matlab* introduced in this chapter, the details and implementation of which are presented in the next chapter has been used for the identification of Mamdani and Singleton fuzzy models

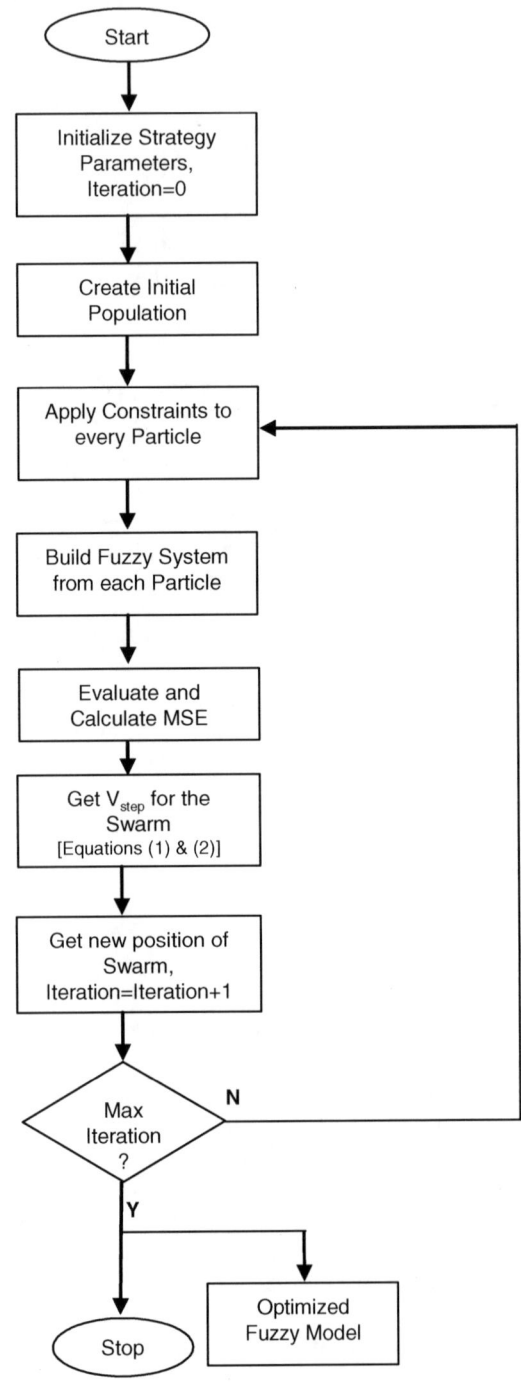

Fig. 7.9. Methodology for fuzzy models identification through PSO

Table 7.3. Input and output variables alongwith their universes of discourse

Input Variables	Universe of Discourse
Temperature (T)	$0 - 50^\circ C$
Temperature Gradient (dT/dt)	$0 - 1(^\circ C/sec)$

Output Variable	Universe of Discourse
Charging Rate (C_t)	$0 - 8C$

from the data. The strategy parameters of PSO algorithm used for the identification of both the models are listed in Table 7.5 and the simulation results obtained are presented in Table 7.5. Centre of Gravity and Weighted Average defuzzification techniques[7] were selected for Mamdani and Singleton fuzzy models respectively.

Simulation results presented in Table 7.5 clearly depict the effectiveness of the proposed methodology and its implementation, as considerable improvement in the performance of fuzzy models was achieved after the complete run of PSO algorithm. More simulation time for Mamdani fuzzy model can be attributed to more complicated, time-consuming defuzzification process.

Table 7.4. Strategy parameters of PSO algorithm for fuzzy models identification

Swarm Size	30
Iterations	2500
c_1	2
c_2	2
w_{start}(Inertia weight at the start of algorithm)	0.9
w_{end}(Inertia weight at the end of algorithm)	0.3
V_{max}	75

Table 7.5. Simulation Results

Experiment	Model	MSE of Fuzzy System Corresponding to Swarm's *gbest*		Simulation time
		After 1st Iteration	After 2500 Iterations	
E1	Mamdani	12.10	0.0488	19.424 hours
E2	Singleton	46.95	0.1118	16.633 hours

7.6 Selection-based PSO with Lifetime Parameter

A new selection-based variant of PSO algorithm, where a new parameter called lifetime is introduced, has been proposed by the authors[19]. Lifetime can be considered analogous to the time-frame generally provided to inefficient workers to improve their performance. In most organizations, non-performing workers are given sufficient opportunities before they are removed from their positions. Then the vacant positions are filled by new incumbents with the hope that they will perform better than their predecessors. In many cases such poor performing workers are able to show performance improvements within the allocated time-frame. In PSO also, a particle with poor fitness value may be able to improve its performance after a few iterations and emerge better than other particles. Angeline, in his work [20] suggested that after every iteration, worst half of the swarm be eliminated and the remaining half be duplicated, with the *pbests* of the destroyed particles retained. Angeline's approach kills diversity and can lead to premature convergence of PSO. If some particles are able to locate a reasonable minimum, the other half of the swarm would be pulled into the basin of same local minimum and this is going to jeopardize the ability of the algorithm to explore large regions of search space, thus preventing it from finding the global minimum. Moreover, since the poor performing particles are destroyed after every iteration, they do not get an opportunity to improve their performance. In the suggested approach, the decision of destroying the worst particle is taken only after fixed number of iterations, defined by lifetime.

The suggested approach is illustrated in Fig. 7.10, where for the purpose of illustration, a two-dimensional system with swarm size of 5 is taken. Fig. 7.10 depicts the situation when the condition defined in (7.15) is satisfied, which implies that the particles have lived their lives and are ready for scrutiny. Assume that at that instance, particle #5 is having the worst fitness value and hence shall be destroyed and new particle is generated in the vicinity of one of the remaining particles chosen through some selection method. In our work, we have used Roulette Wheel Selection [21]. Again it is assumed that particle #2 was chosen. A new particle is generated with co-ordinates of particle #2 and *pbest* of particle #5. This particle is then pushed to place it in the vicinity of the chosen particle by some random push.

$$rem(\frac{Iteration - number}{lifetime}) = 0 \qquad (7.15)$$

The proposed variant was tested on three benchmark functions viz. Rosenbrock, Rastrigin and Griewank. All these three functions have known global minimum equal to zero. The parameters of the PSO were chosen to match the experimental setup adopted by Eberhart and Kennedy in [22]. The values of cognitive and social acceleration parameters, c1 and c2, were kept at 2 for all the experiments. The weight factor was linearly decreased from 0.9 down to 0.4 as the PSO progressed [22]. The velocity-limiting factor, V_{max},

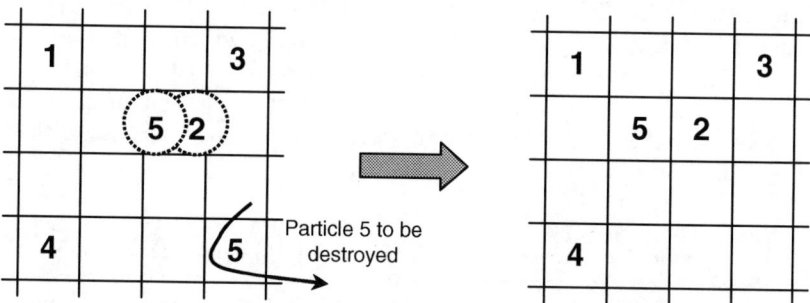

Fig. 7.10. Illustration of the Proposed Approach

was also applied and the values corresponding to each of the functions are shown in Table 7.6 along with the various parameters specific to the proposed model.The particles were initialized asymmetrically [22] and the initialization ranges for the three functions are given in Table 7.7. The three test functions were tested with swarm sizes of 20, 40, and 80. For each of these cases, the dimensions were kept at 10, 20, and 30.

Table 7.6. Parameters for Experiments

Function	$V_{max} = X_{max}$	Lifetime
Rosenbrock	100	5
Rastrigrin	10	5
Griewank	600	5

Table 7.7. Asymmetric Initialization Ranges

Function	Initialization
Rosenbrock	15-30
Rastrigrin	2.56-5.12
Griewank	300-600

Table 7.8, Table 7.9 and Table 7.10 present the results of these experiments that have been reproduced from [19]. Each of the value is the arithmetic mean of the results of fifty experimental runs. The performance of proposed selection based PSO with respect to weighted-PSO as defined by (7.1) and (7.2) is compared by defining performance index, which is the ratio of fitness value

with the weighed-PSO to the fitness value obtained with the new variant. Performance index with value >1 reflects improvement and with value <1 implies deterioration. The comparison of the proposed model with the weighted-PSO reveals that the proposed model performs better for 19 out of 27 instances (slightly more than 70%). The results were especially encouraging for the Rastrigin function.

Table 7.8. Mean Fitness Values for the Rosenbrock function

Population Size	Dimension	Generations	weighted-PSO	New variant	Performance Index
20	10	1000	96.1715	23.1781	4.14924
	20	1500	214.6764	163.77442	1.3108054
	30	2000	316.4468	708.39632	0.4467087
40	10	1000	70.2139	11.97317	5.8642699
	20	1500	180.9671	95.27461	1.8994263
	30	2000	299.7061	208.98225	1.4341223
80	10	1000	36.2945	6.47493	5.6053888
	20	1500	87.2802	112.49097	0.7758863
	30	2000	205.5596	4110.1555	0.0500126

Table 7.9. Mean Fitness Values for the Rastrigrin function

Population Size	Dimension	Generations	weighted-PSO	New variant	Performance Index
20	10	1000	5.5572	3.93286	1.4130175
	20	1500	22.8892	15.91088	1.4385879
	30	2000	47.2941	37.00899	1.2779084
40	10	1000	3.5623	3.04654	1.1692937
	20	1500	16.3504	10.75156	1.5207468
	30	2000	38.525	24.48231	1.5735852
80	10	1000	2.5379	3.10427	0.8175513
	20	1500	13.4263	9.47426	1.4171344
	30	2000	24.0864	19.39997	1.2415689

The methodology for the fuzzy models identification through PSO that was proposed in Section 7.4 was extended to identify Mamdani fuzzy model by following Angeline approach [20] represented in the form of flowchart in Fig. 7.11. Angeline suggested destroying 50% of the worst performing particles after every iteration. Same set of strategy parameters as listed in Table 7.5 were used for this approach. Some simulations were carried out by slightly modifying Angeline approach, where the number of particles to be destroyed

Table 7.10. Mean Fitness Values for the Griewank function

Population Size	Dimension	Generations	weighted-PSO	New variant	Performance Index
20	10	1000	0.0919	0.0213	4.314554
	20	1500	0.0303	0.04918	0.6161041
	30	2000	0.0182	0.42825	0.0424985
40	10	1000	0.0862	0.01483	5.8125421
	20	1500	0.0286	0.02122	1.3477851
	30	2000	0.0127	0.0328	0.3871951
80	10	1000	0.076	0.01724	4.4083527
	20	1500	0.0288	0.02237	1.2874385
	30	2000	0.0128	0.01352	0.9467456

after every iteration was varied. The experiment details and results are listed in Table 7.11.

Table 7.11. Experiment details and results (E3-E6)

Experiment No.	Number of Particles to be destroyed after every iteration	MSE of Fuzzy Model corresponding to Swarm's *gbest* after 2500 iterations
E3	15 (Angeline approach)	0.1219
E4	10	0.047357
E5	5	0.047296
E6	2	0.041834

The convergence plots for these experiments (E1, E3-E6) for identifying Mamdani models are as shown in the Fig. 7.12.

The methodology for the fuzzy models identification through PSO was further extended through incorporating lifetime parameter which is represented as a flowchart in Fig. 7.13. The details of these experiments and simulation results are provided in Table 7.12 and the convergence plots for these experiments are shown in Figure 7.14.

Table 7.12. Experiment details and results (E7-E9)

Experiment No.	Lifetime (Worst performing particle to be destroyed after lifetime)	MSE of Fuzzy Model corresponding to Swarm's *gbest* after 2500 iterations
E7	5	0.041825
E8	10	0.042363
E9	25	0.042395

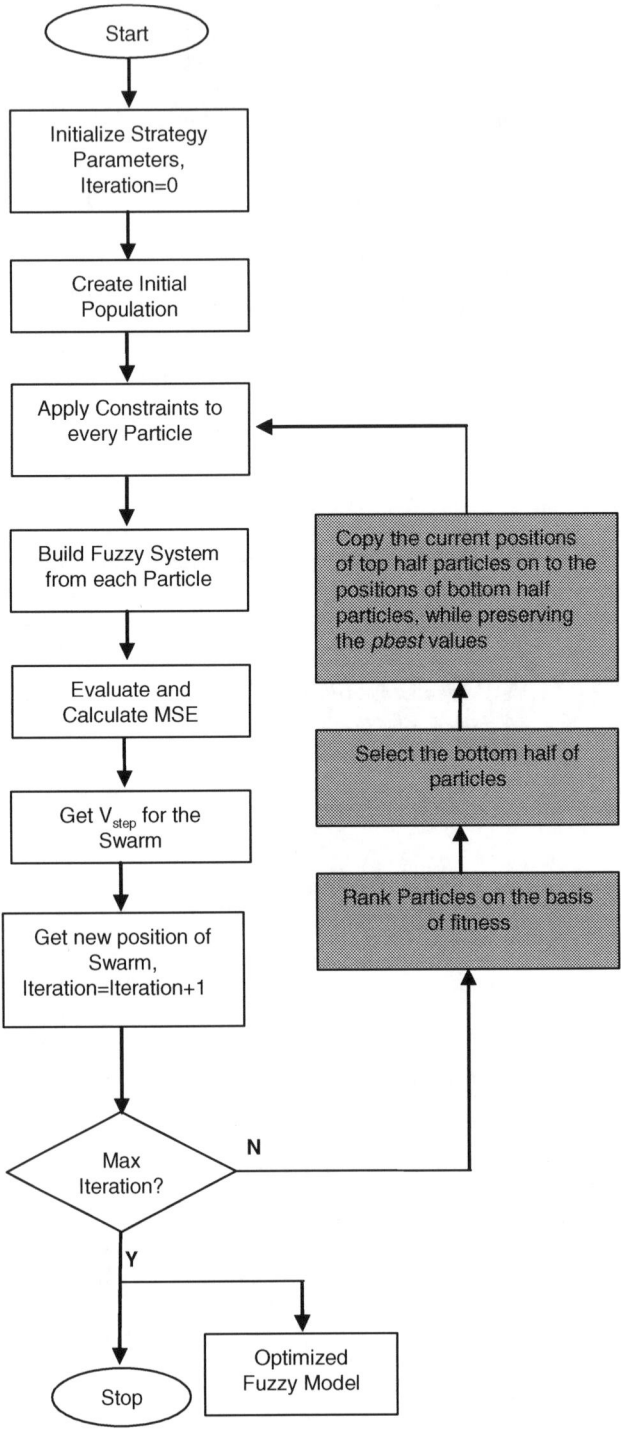

Fig. 7.11. Methodology for fuzzy models identification through PSO with Angeline approach

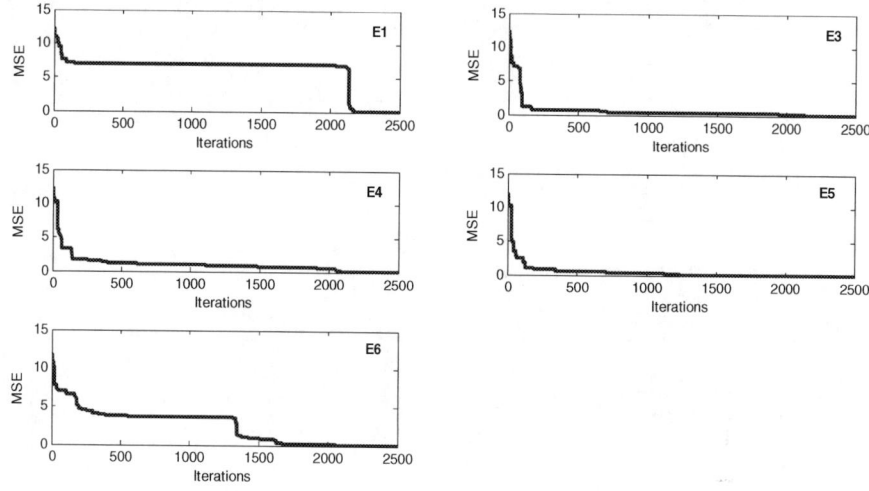

Fig. 7.12. Convergence Plots for experiments E1 and E3-E6

6.1 Analysis of Simulation results

When weighed-PSO was used, as in E1, the end solution quality was good, where a MSE of 0.0488 was obtained after 2500 iterations, but the rate of convergence was very poor. The swarm was able to locate the promising basin (region) only after 2200 iterations.

Followong Angeline approach, as in E3, where 15 (50%) of the particles were destroyed, a very fast convergence was achieved, but the end quality solution was poor. After the complete PSO run, MSE of 0.1219 was reached. Thus, the approach verifies the viewpoint that Angeline approach reduces the diversity and converges quickly to some local minima.

When the number of particles that were destroyed after every iteration was reduced, as in E4-E6, the end results obtained were quite comparable. But for experiments E4 and E5, the convergence rate was good, where the number of particles destroyed after every iteration was 10 (33.33%) and 5 (16.67%) respectively. For E6, where only 2 (6.67%) particles were destroyed after every iteration, the convergence suffered. The results suggest that in order to achieve both good accuracy and convergence simultaneously, less than 50% of particles should be destroyed and this number should not be close to the two extremes: 0%, which represents that no particle is destroyed (experiment E1: poor convergence, good accuracy) and 50%, that represents Angeline approach (experiment E3: good convergence, poor accuracy).

Experiments E7-E9 was carried out after incorporating the lifetime para-meter as discussed in this chapter earlier, where after the lifetime only the worst performing particle is destroyed. For these experiments, for all values of lifetime, the end results were comparable. In E7, for a lifetime of 5, the end

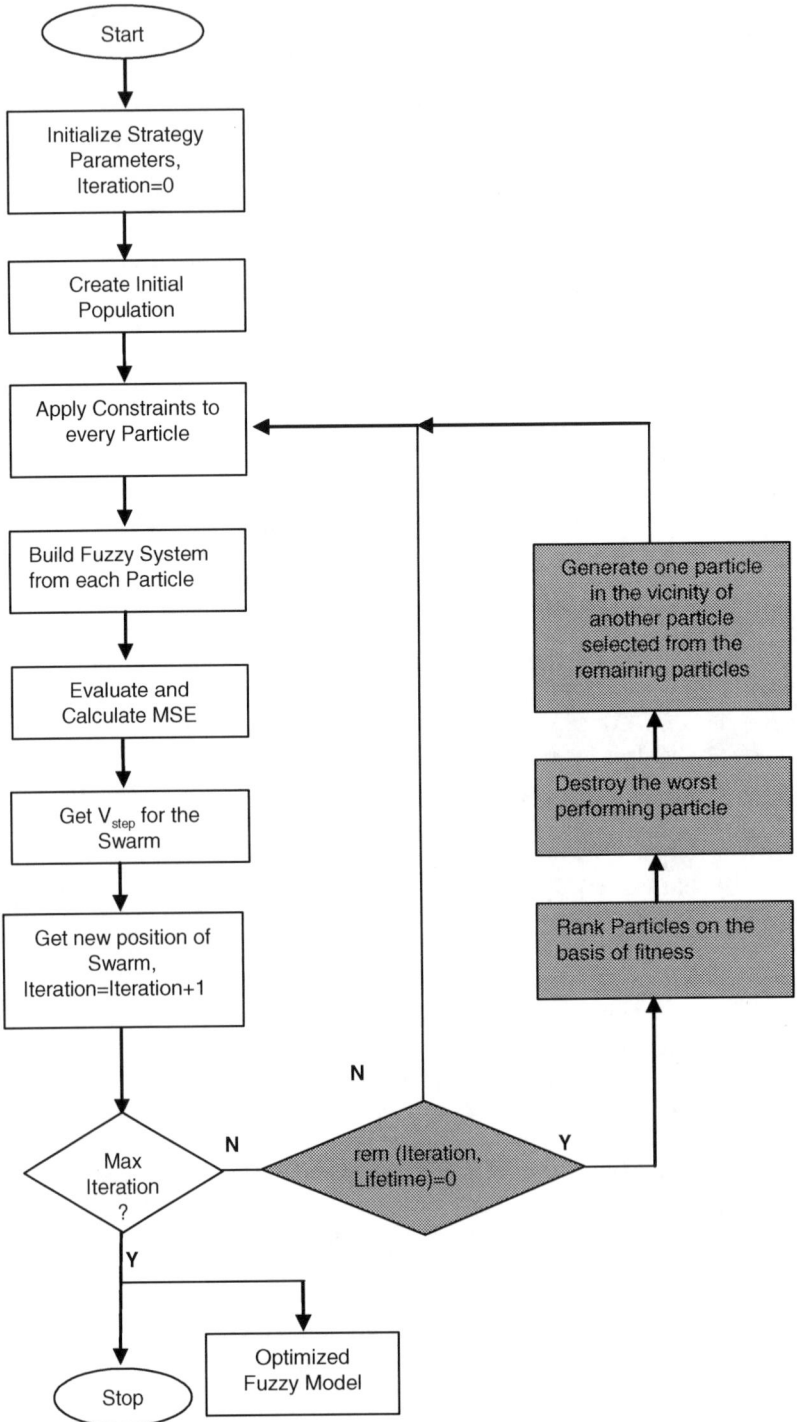

Fig. 7.13. Methodology for fuzzy models identification through PSO with Lifetime parameter

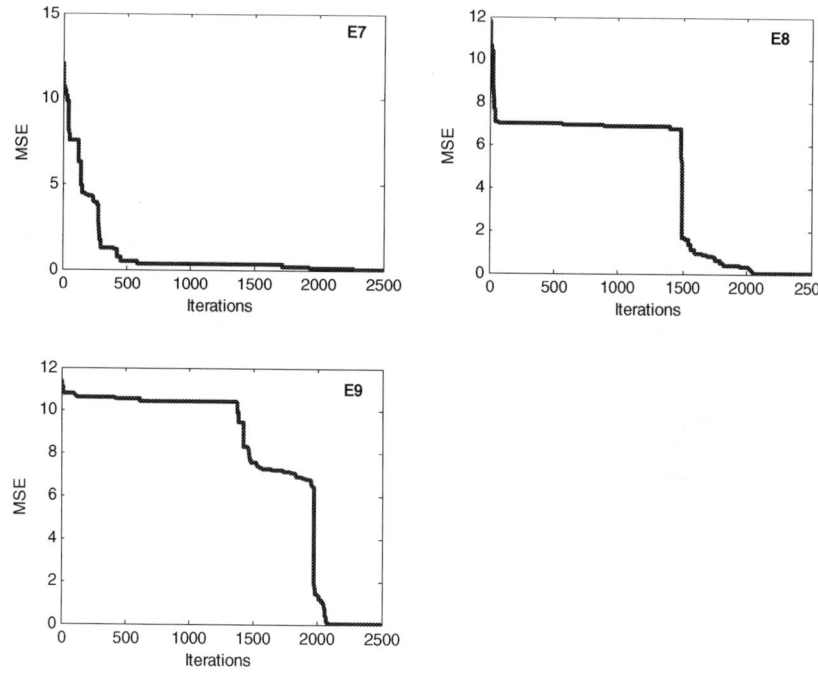

Fig. 7.14. Convergence Plots for experiments E7-E9

solution was not only superior to all the earlier experiments, but a reasonably good convergence was obtained. Another point to note here is that the convergence for E7 was quite superior to E8 and E9, where lifetime values of 10 and 20 respectively was used. Therefore, selection of suitable value of lifetime can help in achieving both; good convergence and accuracy.

7.7 Conclusions and Future Work

In this chapter, the use of PSO algorithm for identification of optimized fuzzy model from the available input-output data is presented. The suggested approach has been implemented as a Matlab toolbox viz. *PSO Fuzzy Modeler for Matlab* that has been presented in the next chapter. All the simulation results reported in this chapter have been obtained using this toolbox and that clearly demonstrates the ability of PSO algorithm for fuzzy models identification. The data from the rapid Ni-Cd battery charger developed by the authors was used for the presentation and validation of the approach.

Simulation results were also reported based on the selection based PSO-variant with lifetime parameter which suggests that the accuracy and convergence can be improved by selecting appropriate values of lifetime parameter and the number of particles to be destroyed.

For all the experiments carried out in this chapter, the swarm size and the number of iterations were kept fixed and it would be worthwhile to investigate the influence of these parameters and trying other PSO variants suggested by different researchers with an objective to achieve good accuracy and convergence. Some of the representative PSO variants are [23][24][25].

Two broad variants of PSO algorithm viz. *gbest* and *lbest* have been developed. The *gbest* model maintains only a single best solution and each particle moves towards its previous best position and towards the best particle in the entire swarm. On the other hand, in the *lbest* model, each particle moves towards its previous best position and also towards the best particle in its restricted neighborhood. It is important to note that the *gbest* model is actually a special case of *lbest* model, when the neighborhood size becomes equal to swarm size. In this chapter, we have used the *gbest* model for fuzzy model identification. The future work would be to incorporate *lbest* model in the methodology used for fuzzy models identification and also to try various neighborhood topologies in *lbest* model.

One of the important future trends is going to focus on augmenting fuzzy modeling with learning and adaptation methodologies based on integration of PSO and other techniques into a hybrid framework.

Another direction for the future work could be applying this methodology for other fields and applications.

References

1. Reynolds CW (1987) Flocks, herds and schools: A distributed behavioral model. Computer Graphics. pp. 25-34.
2. Kennedy J, Eberhart R (2001) Swarm Intelligence. Morgan Kaufmann.
3. Kennedy J, Eberhart R (1995) Particle Swarm Optimization. Proceedings of IEEE Conference on Neural Networks. Perth, Australia. pp. 1942-1948.
4. Eberhart RC, Shi Y (2001) Particle Swarm Optimization: Developments, Applications and Resources. Proceedings of the Congress on Evolutionary Computation, Seoul, Korea. pp. 81-86.
5. Parsopoulos KE, Vrahatis MN (2002) Recent Approaches to Global Optimization Problems through Particle Swarm Optimization. Natural Computing, Kluwer Academic Publishers. pp.235–306.
6. Hellendoorn H, Driankov D (Eds.) (1997) Fuzzy Model Identification - Selected Approaches. Springer-Verlag.
7. Yen J, Langari R (2003) Fuzzy Logic - Intelligence, Control and Information. Pearson Education, Delhi.
8. Bastian A (1996) A Genetic Algorithm for Tuning Membership Functions. Fourth European Congress on Fuzzy and Intelligent Technologies EUFIT(96). Aachen, Germany. pp. 494-498.
9. Carse B, Fogarty TC, Munro A (1996) Evolving Fuzzy Rule-based Controllers using GA. Fuzzy Sets and Systems. pp.273-294.

10. Nelles O (1996) FUREGA–Fuzzy Rule Extraction by GA. Fourth European Congress on Fuzzy and Intelligent Technologies EUFIT(96). Aachen, Germany. pp. 489-493.
11. Nozaki K, Morisawa T, Ishibuchi H (1995) Adjusting Membership Functions in Fuzzy Rule-based Classification Systems. Third European Congress on Fuzzy and Intelligent Technologies, EUFIT(95). Aachen, Germanyvo. pp. 615-619.
12. Shi Y, Eberhart RC, Chen Y (1999) Implementation of Evolutionary Fuzzy Systems. IEEE Transactions on Fuzzy Systems. pp. 109-119.
13. Setnes M, Roubos JA (1999) Transparent Fuzzy Modelling using Clustering and GAs. North American Fuzzy Information Processing Society (NAFIPS) Conference. New York, USA. pp.198-202.
14. Khosla A, Kumar S, Aggarwal KK (2003) Identification of Fuzzy Controller for Rapid Nickel-Cadmium Batteries Charger through Fuzzy c-means Clustering Algorithm. Proceedings of North American Fuzzy Information Processing Society (NAFIPS) Conference. Chicago, USA. pp. 536-539.
15. Melin P, Castillo O (2005) Intelligent Control of a Stepping Motor Drive using an Adaptive Neuro-Fuzzy Inference System. Information Sciences. pp 133-151.
16. Khosla A, Kumar S, Aggarwal KK (2003) Fuzzy Controller for Rapid Nickel-Cadmium Batteries Charger through Adaptive Neuro-Fuzzy Inference System (ANFIS) Architecture. Proceedings of North American Fuzzy Information Processing Society (NAFIPS) Conference. Chicago, USA. pp. 540-544.
17. Khosla A, Kumar S, Aggarwal KK (2002) Design and Development of RFC-10: A Fuzzy Logic Based Rapid Battery Charger for Nickel-Cadmium Batteries. HiPC (High Performance Computing) Workshop on Soft Computing. Bangalore, India. pp. 9-14.
18. Linden D (Editor-in-Chief)(1995) Handbook of Batteries, McGraw Hill Inc.
19. Aggarwal KK, Kumar S, Khosla A, Singh J (2003) Introducing Lifetime Parameter in Selection Based Particle Swarm Optimization for Improved Performance. First Indian International Conference on Artificial Intelligence (IICAI-03). Hyderabad, India. pp. 1175-1181.
20. Angeline PJ (1998) Using selection to Improve Particle Swarm Optimization. Proceedings of IEEE International Congress on Evolutionary Computation. pp. 84-89.
21. Man KF, Tang KS, Kwong S (1999) Genetic Algorithms – Concepts and Designs. Springer-Verlag, London.
22. Shi Y, Eberhart RC (1999) Empirical Study of Particle Swarm Optimization. Proceedings of Congress on Evolutionary Computation. pp. 1945-1950.
23. Eberhart RC, Kennedy J (1995) A New Optimizer Using Particle Swarm Theory. Proceedings of Sixth Symposium on Micro Machine and Human Science. IEEE Service Centre, Piscataway, NJ. pp 39-43.
24. Shi Y, Eberhart RC (2001) Fuzzy Adaptive Particle Swarm Optimization. IEEE International Conference on Evolutionary Computation. pp. 101-106.
25. Xie X-F, Zhang W-J, Yang Z-L (2002) Adaptive Particle Swarm Optimization on Individual Level. 6th International Conference on Signal Processing. pp. 1215-1218.

8

A Matlab Implementation of Swarm Intelligence based Methodology for Identification of Optimized Fuzzy Models

Arun Khosla[1], Shakti Kumar[2], K.K. Aggarwal[3], and Jagatpreet Singh[4]

[1] National Institute of Technology, Jalandhar – 144011, India.
   khoslaak@nitj.ac.in
[2] Haryana Engineering College, Jagadhari – 135003, India.
[3] GGS Indraprastha University, Delhi – 110006, India. kka@ipu.edu
[4] Infosys Technologies Limited, Chennai – 600019, India. jagatpreet@yahoo.com

This chapter presents a Matlab toolbox viz. *PSO Fuzzy Modeler for Matlab*. The toolbox implements the fuzzy model identification procedure using PSO as an optimization engine, which was presented in the previous chapter. This toolbox provides the features to generate Mamdani and Singleton fuzzy models from the available data. The simulation results presented in the previous chapter have been obtained through this toolbox, which is freely distributed on SourceForge.net. SourceForge.net is the world's largest development and download repository of open-source code and applications. This toolbox can serve as a valuable reference to the swarm intelligence community and others and help them in designing fuzzy models for their respective applications quickly.

8.1 Introduction

During the last decade, several scientific languages like Matlab, Mathematica and Modelica have become very popular for both research and educational purposes, but out of these, Matlab is the most popular choice. Matlab is a high-level technical computing language and environment for computation, visualization and programming [1] and is equally popular in academia and industry. Matlab toolbox, which is a collection of Matlab functions, helps in extending its capabilities to solve problems related to some specific domain. Some of the areas in which toolboxes are available include signal processing, image processing, control systems, neural networks, fuzzy logic, wavelets and many others [2]. In this chapter, we present *PSO Fuzzy Modeler for Matlab*, a

A. Khosla et al.: *A Matlab Implementation of Swarm Intelligence based Methodology for Identification of Optimized Fuzzy Models*, Studies in Computational Intelligence (SCI) **26**, 175–184 (2006)

toolbox that implements the swarm intelligence based methodology presented in the previous chapter [3] for the identification of optimized fuzzy models by using PSO algorithm.

This chapter is organized as follows. All the Matlab functions that constitute the toolbox presented in this chapter are introduced in Section 8.2. The role of each of the implemented functions has been described in this section. Section 8.3 presents conclusions and future work directions for enhancing the performance and capabilities of this toolbox.

8.2 PSO Fuzzy Modeler for Matlab

All functions for this toolbox have been developed using Matlab with the Fuzzy Logic toolbox [4] and are listed in Table 8.1. The role of each of the implemented function is explained below in the context of methodology that was presented in the previous chapter to identify fuzzy models using PSO algorithm as an optimization engine [3]. Fig. 8.1 shows the organization of different modules of the Matlab functions that implements the methodology.

Table 8.1. List of Matlab functions

(i)	*RandomParticle*
(ii)	*limitSwarm*
(iii)	*limitParticle*
(iv)	*limitMembershipFunctions*
(v)	*limitRules*
(vi)	*GetFIS*
(vii)	*calculateMSE*

(i) **RandomParticle** – To begin searching for the optimal solution in the search-space, each particle begins from some random location with a velocity that is random both in magnitude and direction. The role of this function is to generate such random particles representing the fuzzy models in the search-space. The particle dimensions equal the search-space dimensions and the number of particles is as defined by the swarm size.

(ii) **limitSwarm** – This function calls another function *limitParticle* for each particle of the swarm.

(iii) **limitParticle** – It is important to always ensure that the particles are confined to the search-space and represent feasible solutions. There are possibilities that during the movement of the swarm, some particles may

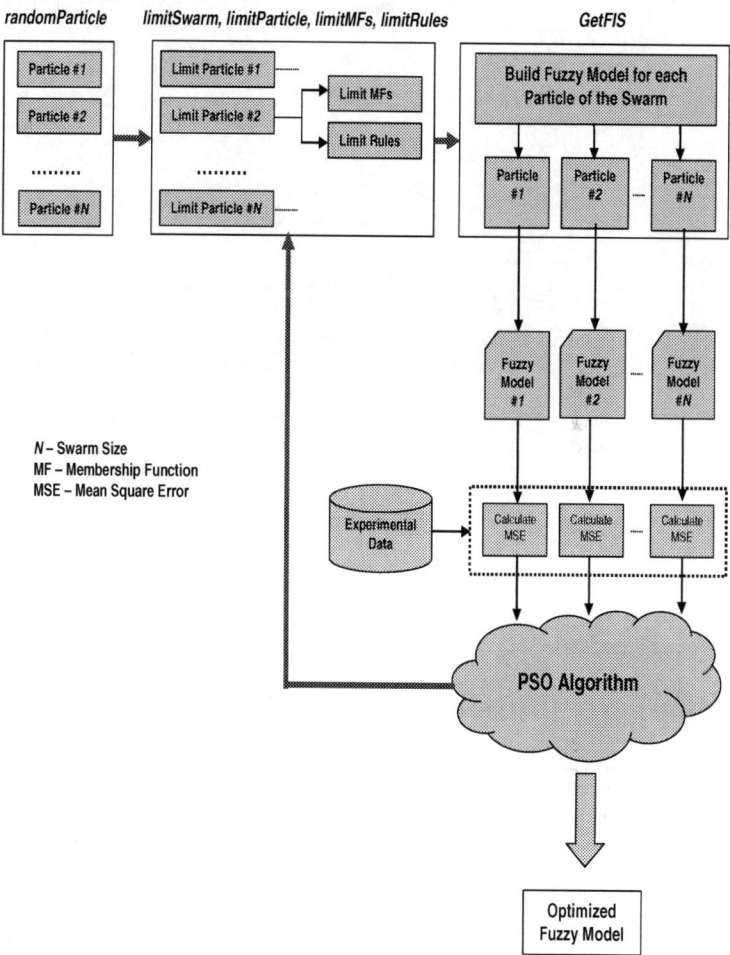

Fig. 8.1. Matlab toolbox modules

move out of the bounds defined by the system constraints. It is therefore
necessary to constrain the exploration to remain inside the valid search-
space. Thus, all the particles in the swarm are scrutinized after every
iteration to ensure that they represent only valid solutions. To illustrate
this, consider that the search-space is three-dimensional represented by
a cube as shown in Fig. 8.2(a). During exploration, some particle may
move out of the search-space as shown in Fig. 8.2(b). All such particles
are required to be brought back to the valid search-space by applying

some limiting mechanism shown in Fig. 8.2(c). The function *limitParticle* is further made up of two functions viz. *limitMembershipFunctions* and *limitRules*.

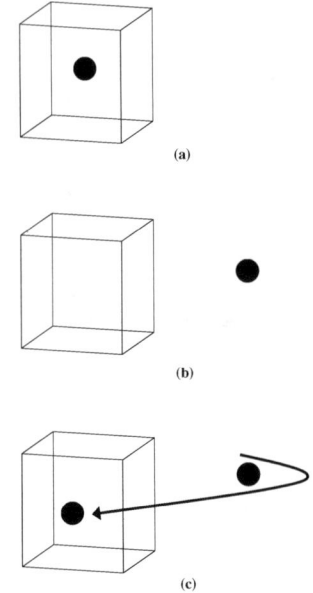

(a)

(b)

(c)

Fig. 8.2. Limiting Mechanism

(iv) **limitMembershipFunctions** – The role of this function is to ensure that membership function parameters for every input and output variable are confined within the respective universe of discourse and at the same time satisfy the constraint defined to ensure overlapping between the adjacent membership functions. Inequality defining these constraints as represented in Fig. 7.5 from the previous chapter [3].
Inequality: $x_{min} \leq x_2^l < x_1^r < x_3^l < x_2^r \leq x_{max}$

(v) **limitRules** – For Mamdani and Singlegon fuzzy models, a fuzzy rule consequent can only refer to one of the membership functions of the output variable. In other words, it can have possible values equal to the number of membership functions of output variable. This limiting can be achieved by using the modulus operator. For example, if there are three membership functions for the output, mod3 of the consequent values for each fuzzy rule is calculated. The rule consequent can be represented as $(x+R)$mod3, where $x=2$ and R is a random number and an integer defined as $1 \leq R \leq 3$.

Since $(x+R)$ can have three possible values of 3, 4 and 5, hence $(x+R)\mathrm{mod}3$ can have three possible values of 0, 1 and 2 that corresponds to the first, second and third membership function of the output variable respectively.

(vi) **GetFIS** – Every particle in the search-space is basically representing a fuzzy model and after every iteration the performance of each fuzzy model is to be worked out to determine the movement of all the particles in the swarm. The role of this function is to generate fuzzy inference system (FIS) from each particle. The Fuzzy Logic Toolbox for Matlab [4] has a structure that can be easily modified. This flexibility has been used for modifying the parameters of fuzzy models through PSO. The FIS structure is the Matlab object that contains all the information about the fuzzy inference system viz. variables names, membership function definitions, rule base etc. [4]. The structure is basically a hierarchy of structures as shown in Fig. 8.3, which can be easily modified by editing its *.fis* text file. For example, the parameters of fuzzy model that are being modified by PSO are represented by the shaded blocks in Fig. 8.3.

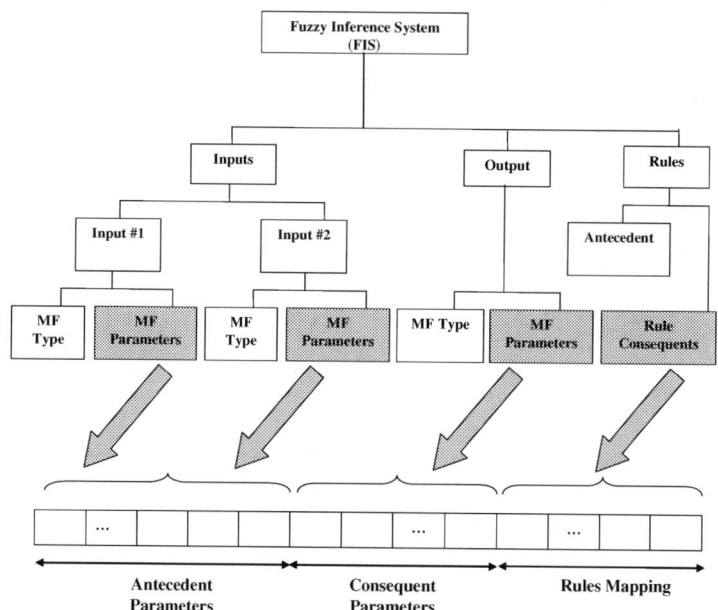

Fig. 8.3. The FIS Structure

(vii) **calculateMSE** – As discussed earlier in the previous chapter, it is imperative to define the fitness/objective function to rate the quality of solutions during the optimization process. This function calculates after every iteration the mean square error (MSE) for each of the fuzzy model represented by each particle of swarm.

It was mentioned earlier in the chapter that *PSO Fuzzy Modeler for Matlab* is hosted on SourceForge.net and is available at [5]. Another toolbox developed by the authors is *PSO Toolbox* (for Matlab), which is also hosted on Source-Forge.net [6]. This toolbox is also a part of *PSO Fuzzy Modeler for Matlab* and implements the PSO algorithm. The organization of various modules of *PSO Fuzzy Modeler for Matlab* is shown in Fig. 8.4.

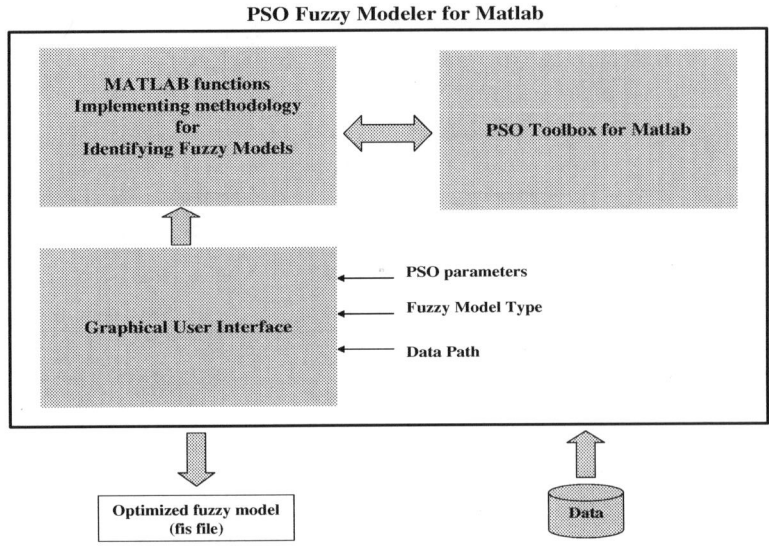

Fig. 8.4. Organization of toolbox modules

A graphical user interface (GUI) has also been designed for the user convenience and is shown in Fig. 8.5. Authors have also proposed a variant of PSO algorithm, by introducing another parameter viz. lifetime[7] and in the previous chapter, we have also presented some simulations results, where this PSO variant has been used for identification of fuzzy models. The GUI for the toolbox, where the proposed variant of PSO algorithm has been used is shown in the Fig. 8.6.

Simulation results can be found Table 7.5 in the previous chapter.

The graphical representation of the input-output battery charger data and surface plots of the corresponding Mamdani and Singleton fuzzy models identified through PSO algorithm are shown in the Fig. 8.7. The plots in Fig. 8.7(b) and Fig. 8.7(c) have been obtained by generating the surface views in

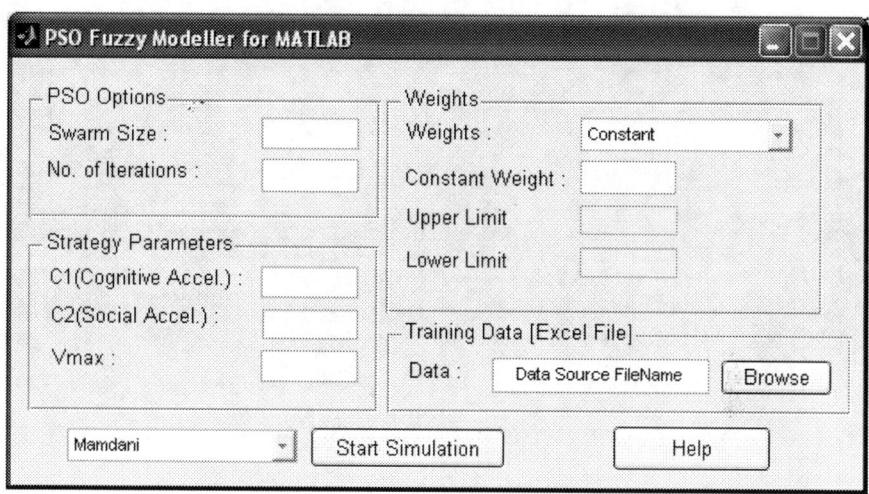

Fig. 8.5. *PSO Fuzzy Modeler for Matlab* GUI

Fuzzy Logic Toolbox for Matlab [4] from the *fis* files created through *PSO Fuzzy Modeler for Matlab.*

8.3 Conclusions and Future Work Directions

In this chapter, a Matlab toolbox viz. *PSO Fuzzy Modeler for MATLAB* is presented. The toolbox implements the methodology based on PSO algorithm for the identification of optimized fuzzy models from the available input-output data. This GUI based toolbox, which is hosted on SourceForge.net as an open source initiative, has the capabilities to generate Mamdani and Singleton fuzzy models from the available data and is going to help the designers build fuzzy systems from their data quickly. The data from the rapid Ni-Cd battery charger developed by the authors was used for the presentation and validation of the approach. Simulation results presented in the previous chapter have been generated through this toolbox and is clearly indicative of the suggested methodology abilities and its implementation as a Matlab toolbox.

The parallel nature of evolutionary algorithms requires lot of computational efforts, which is evident from the simulation time reported in Table

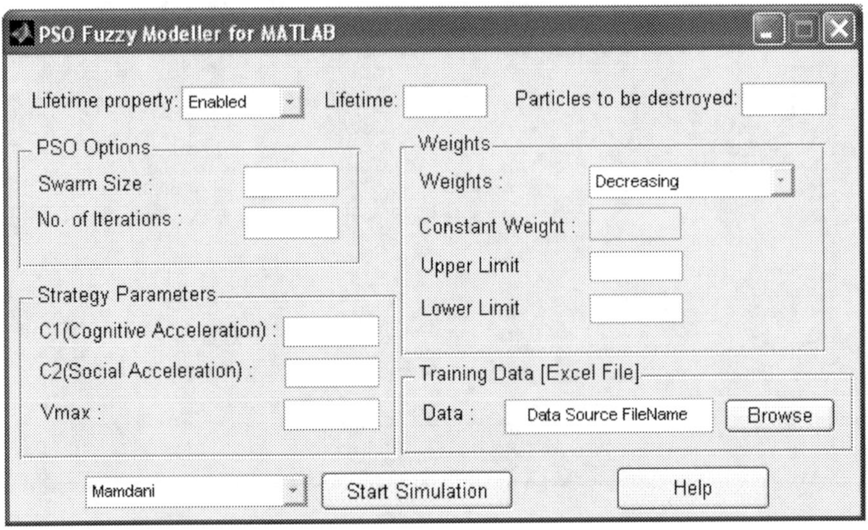

Fig. 8.6. *PSO Fuzzy Modeler for Matlab* GUI implementing PSO with lifetime parameter

YYYY.5 of the previous chapter for the given data. The computer time is directly proportional to the complexity of the problem under consideration and for a practical system, the simulation time may run into many days or even months. Thus, the use of high performance computing resources becomes a key to obtaining useful answers in acceptable amounts of time. This can be achieved through cluster computing by employing off-the-shelf hardware and software systems. A cluster is a group of independent computers working as a single, integrated computing resource. The cluster computing has become the paradigm of choice for executing large-scale science, engineering and commercial applications. One of the possible approaches for building such a solution is through the use of Cornell Multitask Toolbox (CMTM) [8][9], developed by Cornell Theory Centre to enable Matlab for parallel processing through MPI (message passing interface) paradigm. Another alternative is to use proprietary Matlab toolboxes viz. Distributed Computing Toolbox [10] and Matlab Distributed Computing Engine [11] recently released by Mathworks. These toolboxes helps in developing distributed applications in Matlab that can run on a computer cluster. For both these implementations the toolbox presented in this paper shall be required to be modified so as to run on the cluster.

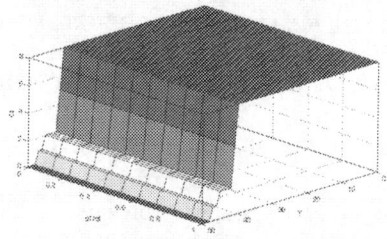

(a) Surface plot generated from the input-output data

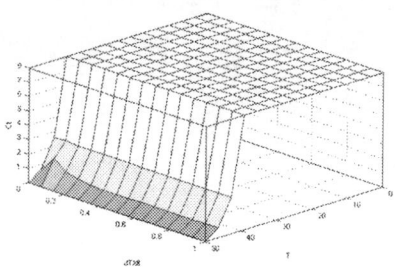

(b) Surface plot for the identified Mamdani fuzzy model

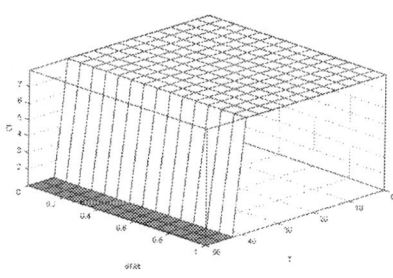

(c) Surface plot for the identified Singleton fuzzy model

Fig. 8.7. Graphical representation

Two broad variants of PSO algorithm were developed: one with a global neighborhood called *gbest* model and the other with local neighborhood known as *lbest* model [12]. The *gbest* model maintains only a single best solution and each particle moves towards its previous best position and towards the best particle in the whole swarm. The best particle acts as an attractor, pulling all the particles towards it. In the *lbest* model, each particle moves towards its previous best position and also towards the best particle in its restricted neighborhood and thus maintains multiple attractors. Although the *gbest* model is most commonly used, it is vulnerable to premature convergence. The toolbox presented in this paper is created around the *gbest* model. One of the future directions could be to incorporate in the toolbox the *lbest* model and other PSO variants suggested [13][14][15].

The toolbox in true sense can be called an open-source only if both the toolbox and the platform on which the toolbox runs should be free. Since the Matlab environment is commercial, this may become a hindrance in exchanging ideas and further improvements in the toolbox design from people who doesn't use Matlab. One of the important tasks for future would be to develop such tools/applications in Java or other high level languages so as to make them platform independent for wider usage, exchange and improvements.

References

1. http://www.mathworks.com
2. http://www.mathworks.com/products/product_listing/index.html
3. Khosla A, Kumar S, Aggarwal KK, Singh J (2006) Particle Swarm for Fuzzy Models Identification. In: Nadia Nedjah, Luiza Mourelle (Eds.) Swarm Intelligent Systems. Springer-Verlag, Berlin. this book.
4. Jang JSR, Gulley N (1995) Fuzzy Logic Toolbox User's Guide. The Mathworks Inc., USA.
5. PSO Fuzzy Modeler for Matlab
   http://sourceforge.net/projects/fuzzymodeler
6. PSO Toolbox (for Matlab)
   http://sourceforge.net/projects/psotoolbox
7. Aggarwal KK, Kumar S, Khosla A, Singh J (2003) Introducing Lifetime Parameter in Selection based Particle Swarm Optimization for Improved Performance. First Indian International Conference on Artificial Intelligence, Hyderabad, India. pp. 1175–1181.
8. www.tc.cornell.edu/Services/Software/CMTM
9. Bekas C, Kokiopoulou E, Gallopoulos E (2005) The design of a distributed MATLAB-based environment for computing pseusospectra. Future Generation Computer Systems 21:930–941.
10. http://www.mathworks.com/products/distribtb
11. http://www.mathworks.com/products/distriben
12. Parsopoulos KE, Vrahatis MN (2002) Recent Approaches to Global Optimization Problems through Particle Swarm Optimization. Natural Computing, Kluwer Academic Publishers. pp.235–306.
13. Eberhart RC, Kennedy J (1995) A New Optimizer Using Particle Swarm Theory. Proceedings Sixth Symposium on Micro Machine and Human Science. pp. 39–43.
14. Shi Y, Eberhart RC (2001) Fuzzy Adaptive Particle Swarm Optimization. IEEE International Conference on Evolutionary Computation. pp. 101–106.
15. Xiao-Feng Xie, Wen-Jun Zhang, Zhi-Lian Yang (2002) Adaptive Particle Swarm Optimization on Individual Level. International Conference on Signal Processing (ICSP 2002). pp. 1215–1218.

Author Index